OPTIMAL CONTROL *and* DIFFERENTIAL EQUATIONS

Academic Press Rapid Manuscript Reproduction

Proceedings of the
Conference on Optimal Control and Differential Equations
held at the University of Oklahoma
Norman, Oklahoma, March 24–27, 1977

OPTIMAL CONTROL *and* DIFFERENTIAL EQUATIONS

Edited by

A. B. SCHWARZKOPF
WALTER G. KELLEY
STANLEY B. ELIASON

Department of Mathematics
The University of Oklahoma
Norman, Oklahoma

ACADEMIC PRESS New York San Francisco London 1978

A Subsidiary of Harcourt Brace Jovanovich, Publishers

ACADEMIC PRESS, INC.
111 Fifth Avenue, New York, New York 10003

United Kingdom Edition published by
ACADEMIC PRESS, INC. (LONDON) LTD.
24/28 Oval Road, London NW1 7DX

Library of Congress Cataloging in Publication Data

Conference on Optimal Control and Differential Equations, University of Oklahoma, 1977.
Optimal control and differential equations.

1. Control theory—Congresses. 2. Mathematical optimization—Congresses. 3. Differential equations—Congresses. I. Schwarzkopf, Albert B. II. Kelley, Walter G. III. Eliason, Stanley B. IV. Title.
QA402.3.C577 1977 629.8'312 78-6274
ISBN 0-12-632250-3

PRINTED IN THE UNITED STATES OF AMERICA

IN MEMORIAM

It has been the writer's good fortune to know W. T. Reid since 1936, when he was one of the bright young members of the Department of Mathematics at the University of Chicago. We were colleagues in the System Evaluation Department of the Sandia Corporation during portions of 1952 and 1953 and in the Department of Mathematics at The University of Oklahoma from 1964 until 1976.

His enthusiasm and meticulous grasp both of theories and of essential details were shown in the classroom, but especially in seminars and conferences on differential equations and variational theory. He believed in the highest standards for mathematical education and yet was patient with those of limited ability who were struggling to follow. He shared generously the breadth and depth of his knowledge with anyone who sought his counsel. W. T. and Idalia have been the best of companions on social occasions.

The death of this good man, this cherished friend, on October 14, 1977, represents a profound personal loss to his many former students and associates and is a material loss to the mathematical community.

George M. Ewing

Contents

List of Contributors

Numbers in parentheses indicate the page on which the authors' contributions begin.

LEONARD D. BERKOVITZ (107), Department of Mathematics, Purdue University, West Lafayette, IN 47907

GARRET J. ETGEN (245), Department of Mathematics, University of Houston, Houston, TX 77004

GEORGE M. EWING (87), Department of Mathematics, The University of Oklahoma, Norman, OK 73019

PHILIP HARTMAN (293), Department of Mathematics, The Johns Hopkins University, Baltimore, MD 21218

MAGNUS R. HESTENES (165), Department of Mathematics, University of California, Los Angeles, Los Angeles, CA 90024

M. Q. JACOBS (151), Department of Mathematics, University of Missouri-Columbia, Columbia, MO 65201

VADIM KOMKOV (51), Department of Mathematics, Texas Tech University, Lubbock, TX 79409

ROGER T. LEWIS (245), Department of Mathematics, University of Alabama, Birmingham, AL 35294

E. J. McSHANE (3), Department of Mathematics, The University of Virginia, Charlottesville, VA 22903

JAMES S. MULDOWNEY (317), Department of Mathematics, The University of Alberta, Edmonton, Alberta, Canada

W. C. PICKEL (151), Department of Mathematics, University of Missouri-Columbia, Columbia, MO 65201

EVERETT PITCHER (223), Department of Mathematics, Lehigh University, Bethlehem, PA 18015

RAYMOND REDHEFFER (277), Department of Mathematics, University of California, Los Angeles, Los Angeles, CA 90024

W. T. REID* (189), Austin, TX 78701

GEORGE SEIFERT (331), Department of Mathematics, Iowa State University, Ames, IA 50010

JACK WARGA (131), Department of Mathematics, Northeastern University, Boston, MA 02115

*Deceased.

Preface

The articles in this volume were all presented at the Conference on Optimal Control and Differential Equations held at The University of Oklahoma in Norman, March 24–27, 1977. The occasion for this conference was the retirement of Professors W. T. Reid and George M. Ewing from the faculty of The University of Oklahoma. Since their retirement also signaled the passing of a generation of mathematicians who made fundamental advances in the calculus of variations and related problems in differential equations, it seemed appropriate to mark this occasion with a conference that would attempt to assess the present state of mathematical knowledge in these areas and suggest directions for new research efforts.

We invited the authors appearing in this volume to present talks on their own fields of expertise describing the field, rather than presenting new research. The conference itself attracted nearly one hundred participants, and was made a success through the efforts of many people. We would like to acknowledge especially the financial support of the Army Research Office, Durham, Grant No. DAAG29-77-M-0059. We also appreciate the encouragement and financial support of Dean Paige Mulhollan of the College of Arts and Sciences; Dean Gordon Atkinson of the Graduate College; and Dr. Gene Levy, Chairman of the Department of Mathematics, all at The University of Oklahoma.

We finally acknowledge the special talents of Ms. Trish Abolins for her expert work as a copy editor and technical typist for the preparation of this volume.

Optimal Control

THE CALCULUS OF VARIATIONS FROM THE BEGINNING THROUGH OPTIMAL CONTROL THEORY

E. J. McShane

University of Virginia
Charlottesville, Virginia

Before I begin this talk, I would like to sketch briefly what I plan to do. I hope to speak of some of the important stages of the development of the calculus of variations, with a disproportionately large part of the hour allotted to recent developments. But I have no intention of listing important discoveries with their dates. Rather, I shall try to say something of the underlying patterns of thought at each stage, and to comment on the change in that pattern produced by each of the new ideas. It may seem that I am deriding our predecessors for not having seen at once all that we have learned. I have no such intention. We must all do our thinking on the foundation of what we already know. It is hard to assimilate a genuinely new idea, and even harder to realize that ideas we have earlier acquired have become obsolete.

ISBN 0-12-632250-3

Preparing this talk has forced me to formulate with at least some pretension to clarity what is meant by the calculus of variations. There is no universal agreement on the definition of the subject, and I have gradually come to the conclusion that part of the reason is that there are at least two related but different sets of ideas that are often brought together under the same name. The first set might be called the theory of extrema. A functional is defined on some class of functions; the problem is to find a function in the given class that minimizes or maximizes the functional on that class. If this theory of extrema is included in the calculus of variations, Caratheodory may be justified in asserting that the first problem in the calculus of variations was that of finding a curve of given length that joins the ends of a line segment, and together with that segment encloses the greatest possible area. This was solved, according to Caratheodory, by Pappus, in about 290 A.D.

The second set of ideas is concerned with functionals on linear topological spaces, ususally function spaces, and constitutes a part of a differential calculus on such spaces. The central problem in this part of the theory is that of finding stationary points of functionals; that is, points at which the directional derivatives in all directions exist and are all 0 . Since such points are characterized by means of investigating the effect on the functional produced by small variations of the function which is the independent variable, this study of stationary points can reasonably be called the calculus of variations.

The two sets of ideas both have important applications, but to different problems. At one extreme we have those problems such as the isoperimetric problem of Pappus just mentioned, and more recently problems in which a function is to be found that produces a best possible result in some sense, such as propelling an airplane between given points with least expenditure of fuel. At the other extreme we have situations in which the presence or absence of a maximum or minimum is irrelevant; only the consequences of stationarity matter. These consequences often include the satisfaction of a set of differential equations. According to what is misnamed "the principle of least action," the motion of a set of particles follows a time-development for which a certain integral, called the "action," is stationary. The function for which the action is stationary is the one for which the classical equations of motion are satisfied, and the satisfaction of those equations is all that we want.

In between these two extremes we have the problems of relative extrema. Let us say that a function y is in the weak ε-neighborhood of another function y_0 if there is a homeomorphism between their graphs such that at corresponding points, the values of y and y_0 differ by less than ε, and so do the values of their derivatives. The function y is in the strong ε-neighborhood of y_0 if this holds with the reference to the derivatives deleted. A functional has a weak (strong) relative minimum at y_0 if for some positive ε, the functional has at y_0 its least value on the set of all those y in the domain of the functional that are in the weak (strong) ε-neighborhood of y_0. These concepts have some applications, related to stable and unstable equilibrium; but I have a strong suspicion that relative maxima and minima were usually studied, not because they were really wanted, but because available theory did not permit the study of absolute maxima and minima.

For lack of time I shall say little about the second set of ideas, based on stationarity. This means that I shall disregard some important pure mathematics and some important applications. I have mentioned that the principle of least action is of this type. So too is Hamilton's study of optics and its extension into calculus of variations by Jacobi. So is all the mathematics of quantum theory that is based on a Hamiltonian. So, too, is Marston Morse's theory of the calculus of variations in the large. I shall choose for my principal subject the development of the first set of ideas, that I have called the theory of extrema.

In the eighteenth century the distinction between the two sets of ideas was hardly noticed. If it could be shown that any curve that minimized some functional had to satisfy a certain condition, and a curve could be found that did satisfy that condition, it was accepted without comment that that curve did furnish the minimum. Nor has such a feeling quite disappeared. On page 16 of the book by Gelfand and Fomin (English translation) we read: "In fact, the existence of an extremum is often clear from the physical or geometric meaning of the problem, e.g., in the brachistochrone problem, the problem concerning the shortest distance between two points, etc. If in such a case there exists only one extremal satisfying the boundary conditions of the problem, this extremal must perforce be the curve for which the extremum is achieved." I disagree with this on three counts. First, if the calculus of variations is mathematics, our conclusions must be deducible logically from the hypotheses, with no use of anything that is "clear from the physical meaning" — even if anything is ever that clear in physics. Second, if the mathematical expression is meant to be a model of a physical situation, we are not entitled to unshakeable confidence that the model we have chosen is perfect in all details; rather, we should keep in mind that a mathematical model of a physical system is necessarily a simplification and idealization. Third, the principle as stated is untrustworthy. For example, if A and B are two points in the upper half-plane there always exists a curve joining them such that the surface of revolution obtained by rotating it about the x-axis has least area. If A and B

are properly located, there is just one extremal that joins them, and it does not furnish the least area. (See G. A. Bliss, "Calculus of Variations," p. 116.)

In the early eighteenth century the necessary conditions for a minimum in various specific problems were found by ingenious devices, usually involving replacing a short arc of the curve by another short arc with the same ends. In 1760, Lagrange unified these special solutions by means of the idea of a variation. Suppose that a function $x \to y(x)$ $(x_0 \leqq x \leqq x_1)$ minimizes a functional $J(x(\cdot))$ in a certain class K of functions. Suppose further that we can find a family of functions y_α $(-b < \alpha < b)$ such that for each α in $(-b,b)$ the function $x \to y_\alpha(x)$ $(x_{0,\alpha} < x < x_{1,\alpha})$ is in the given class K. Then the derivative at $\alpha = 0$ of the function $J(y_\alpha(\cdot))$, if it exists, must be 0. The function

$$x \to \eta(x) = \partial y_\alpha(x)/\partial\alpha \qquad (\alpha = 0)$$

is often called a variation of y; Lagrange used the term "variation" and the symbol δy for the product of this by $d\alpha$ The variation of the functional, which is the derivative of $J(y_\alpha(\cdot))$ at $\alpha = 0$, is the directional derivative of J in the direction η. In many interesting cases its vanishing is equivalent to the satisfaction of a certain differential equation; this is the Euler-Lagrange equation.

For the purposes of mechanics the goal had now been reached. The Euler-Lagrange equation permitted the introduction of general coordinate systems, and the concept of stationary curve unified the whole theory of classical mechanics, as Lagrange showed in his masterful work. But it was a mental confusion, consistent with the somewhat uncritical ideas of the period, to think that any stationary curve would certainly furnish a maximum or a minimum, as wished. In his "Principia,"

Isaac Newton had discussed the problem of finding a surface of revolution with assigned base and altitude that minimized a functional that Newton thought represented the drag when the body is moved through a fluid. Legendre published his necessary condition for a minimum in 1786, a century later; but in 1788, he published another paper, entitled "Mémoire sur la manière de distinguer les maxima des minima dans le calcul de variations," in which he pointed out that a curve could satisfy the Euler-Lagrange equation for the integral expressing the Newtonian resistance and still not give the surface of least resistance. The most interesting feature of his proof is that he showed that the Weierstrass condition for a minimum was not satisfied — and Weierstrass was born until twenty-seven years later. This work must not have had the immediate effect that it deserved. Mathematicians continued to act as though the only feature of importance was the satisfaction of the condition for stationarity. More than two decades later Robert Woodhouse, F. R. S., a Fellow of Caius College, Cambridge, published a book entitled "Treatise on Isoperimetrical Problems and the Calculus of Variations," (1810), in which Legendre is not mentioned. In this book, Woodhouse poses the problem of maximizing the integral

$$\int [d^2y/dx^2]^2 dx ,$$

the class of curves not being clearly specified. By use of variations he came to the conclusion that the maximum is provided by the line segment joining the end-points. Had he used Legendre's results he would have recognized the falsity of his conclusion. But even without having read Legendre, he should have noticed that unless the end-points coincide, no maximum can exist, and the line segment gives to the integral the value 0 , an obvious minimum.

The guiding principle during the eighteenth century and more than half the nineteenth seemed to be that if a minimizing function is sought for some functional, then by inventing more and more necessary conditions for a minimum we can feel steadily more confident that a function that passes all the tests is in fact the minimizing function sought. The first necessary condition was stationarity, established when the curve being tested can be varied in arbitrary directions. The next in order of time was the Legendre condition, still in the domain of Lagrange-type variations, and in fact needing only variations that leave the function unchanged outside a small interval. Next came the condition of Jacobi. Like that of Legendre, it expressed the fact that for a minimum all directional second derivatives (second variations) must be nonnegative; but unlike Legendre's it required the variation of the function along long intervals. Next came the necessary condition of Weierstrass. Unlike the others, it cannot be established by means of Lagrange-type variations or directional derivatives. The function being tested is compared with other functions near it in position but widely different in derivative. That is to say, the Weierstrass condition is necessary for a strong relative minimum, not for a weak one.

But Weierstrass made a more significant contribution than the discovery of a new necessary condition. For unconditioned problems, in which the minimum of an integral

$$\int_{x_0}^{x_1} f(x,y(x),y'(x))dx$$

is sought in the class of all sufficiently well-behaved functions with assigned end-values, he was able to prove that when a function $y(\cdot)$ satisfies conditions that are slight strengthenings of the four known necessary conditions, it will provide a strong relative minimum for the integral. Now, at last, instead of feeling confident without conclusive proof that a curve gave a minimum to the integral, we could feel certain that it gave a kind of minimum — not indeed the absolute minimum that we were seeking, but at least a strong relative minimum.

This was truly a great step forward in the theory of the calculus of variations. (It might help to promote humility among us workers in that field if we notice that in his biography of Weierstrass in "Men of Mathematics," E. T. Bell does not even mention that Weierstrass wrote on the calculus of variations.) But it had a psychological drawback. Like the ideas introduced by Lagrange a century earlier, the means used by Weierstrass were so highly esteemed that they became ends in themselves. No matter what the calculus of variations was formally stated to be, in the hands of many of its workers it became a procedure of proving in each new type of problem some analogues of the necessary conditions of Euler and Lagrange, of Legendre, of Jacobi and of Weierstrass, and then of proving a sufficiency theorem of some sort. This is not astonishing. When I studied calculus, a mere fifty-five years ago, the theory of maxima and minima consisted of finding points at which the derivative of a function is 0 and then looking at the value of the second derivative. I learned a needed lesson years later, when for quite practical reasons I needed to find the absolute minimum of a function, and discovered that setting the derivative equal to 0 located the maximum; the minimum that I needed was at an end-point, where the derivative was not 0.

Beginners in calculus today are taught a better method of finding minima of functions f on a closed interval $[a,b]$. First it is shown (or at least asserted in an authoritative tone of voice) that a minimum exists. Next, conditions are found that must be satisfied at the point x_0 at which f is minimum; either x_0 is a or b, or the derivative exists at x_0 and is 0, or the derivative does not exist at x_0. In many problems these necessary conditions rule out all but a few values of x. One of these gives f its least value; which one can be determined by calculating f at these points. A similar method could be used to find the absolute minimum of a functional provided that first an existence theorem is proved, and then necessary conditions are found that have to be satisfied at the minimizing function. If these necessary conditions rule out all but a few functions, calculating the corresponding values of the functional will permit us to find which one furnishes the absolute minimum.

Early in this century David Hilbert proved an existence theorem for certain unconditioned problems. Later, Leonida Tonelli showed that if an integrand $f(x,y,y')$ is convex as a function of y', its integral is lower semi-continuous; if a sequence of functions $y_1, y_2, \ldots$ tends in the strong topology to a limit function y_0, the limit inferior of the integral along y_n is at least equal to the integral along y_0. This, with some other very reasonable hypotheses, gave excellent existence theorems for unconditioned problems. But when applied to conditioned problems, such as isoperimetric problems and Bolza problems, it produced no results of interest. These conditioned problems are neither new nor artificial. As to newness, Forsythe asserts that the word "isoperimetric" was first used in the early fifth century, by Bishop Synesius. As to artificiality, the engineering problems that led the Russian mathematicians to devise the modern form of control theory are almost invariably conditioned problems. Conditioned problems were unmanageable until L. C. Young invented what he called generalized curves.

For those of us who have not encountered generalized curves, a bit of explanation might be helpful. Suppose that we wish to minimize the integral

$$\int_0^1 [y^2 + (y'^2 - 1)^2]dx \tag{1}$$

in the class of absolutely continuous functions $y(\cdot)$ on $[0,1]$. If we divide $[0,1]$ into $2n$ intervals of equal length and define y_n to be the function with $y_n(0) = 0$ and $y' = 1$ and $y' = -1$ on alternate subintervals, the graph of y_n is a saw-tooth polygon, and the integral has value $1/12n^2$. So the lower bound of the integral is 0. But y_n tends uniformly to the 0 function, for which the integral (1) has value 1. We need a different approach. Let us plot each value of y_n' on a u-axis. If we select a subinterval of $[0,1]$ and choose an x at random in it, there is a certain probability that $y_n'(x) = 1$, and this probability tends to $\frac{1}{2}$ as n increases; and likewise the probability that $y_n'(x) = -1$ tends to $\frac{1}{2}$ as n increases. So we construct a new kind of object. Instead of having a number $y'(x)$ associated with each x in $[0,1]$, it has a probability distribution P_x that assigns probability $\frac{1}{2}$ to each of the numbers $1,-1$ and the probability 0 to the rest of the real number system. This is what we would have in the unrealizable situation that on every subinterval of $[0,1]$, $y'(x)$ were $+1$ half the time and -1 half the time. This distribution takes the place of the single number $y'(x)$, which can be thought of as a distribution in which probability 1 is assigned to the number $y'(x)$ and 0 to the rest of the real numbers. Thus instead of having $y_0(x_1)$ equal to the integral of $y_0'(x)$ from 0 to x_1 $(0 \leqq x_1 \leqq 1)$, we have

$$y_0(x_1) = \int_0^{x_1} \{\int_R uP_x(du)\}dx = 0 ;$$

and likewise the replacement for the integral (1) is

$$\int_0^1 \{\int_R [y_0{}^2 + (u^2 - 1)^2]P_x(du)\}dx ,$$

which has the value 0 . Young's generalized curves are objects of this new kind. A generalized curve can be thought of as a pair $((y(x),P_x) : a \leqq x \leqq b)$ in which the y is a function on $[a,b]$, and for each x in $[a,b]$, P_x is a probability distribution on R , and

$$y(x_1) = \int_a^{x_1} \{\int_R uP_x(du)\}dx , \quad (a \leqq x_1 \leqq b).$$

(This is close to Young's original formulation; in his book "Calculus of Variations and Optimal Control Theory" he prefers to regard a generalized curve as a functional on a class of integrands, somewhat like Schwartz distributions.) This can be generalized at once to higher dimensions.

The space of generalized curves can be topologized by defining the statement that a sequence of generalized curves $((y_n(x),P_{n,x}) : a_n \leqq x \leqq b_n)$ tends to a generalized curve $((y_0(x),P_{0,x}) : a_0 \leqq x \leqq b_0)$ if and only if

$$\lim_{n\to\infty} \int_{a_n}^{b_n} \{\int_R \phi(x,y_n(x),u)P_{n,x}(du)\}dx$$

$$= \int_{a_0}^{b_0} \{\int_R \phi(x,y_0(x),u)P_{0,x}(du)\}dx$$

for every continuous function ϕ that vanishes outside a bounded set. The extension to higher dimensions is obvious. The remarkable fact is that with this topology, the space of generalized curves has sufficiently strong compactness properties so that for a large class of problems, not merely unconditioned problems, a minimizing generalized curve can be found for the integral under consideration. Young applied this in 1937 to unconditioned problems. In 1940, I published a sequence of three papers in which, for problems of Bolza in parametric form, it was shown first that under weak hypotheses a minimizing generalized curve exists; second, that it satisfies necessary conditions that are generalizations of the Euler-Lagrange, Legendre and Weierstrass conditions; and third, that under some extra hypotheses, the minimizing generalized curve has each probability measure P_x concentrated at a single point, so that it is in fact an ordinary curve in another notation.

This set of papers burst on the mathematical world with all the éclat of a butterfly's hiccough. The reaction of mathematicians was like that of the little boy who wrote his grandmother: "Thank you for the book about penguins. It taught me more than I wanted to know about penguins." Because it provided a means of finding extrema analogous to today's method of treating minima in calculus, it extended to mathematicians the privilege of forgetting about semi-continuity and about sufficiency theorems. But it was superfluous. Without it, almost all of them had already forgotten about semi-continuity and about sufficiency theorems. And they were justified. The problem of Bolza was the most general of the single-integral problems of the calculus of variations. Its mastery gave us the power to answer many deep and complicated questions that no one was asking. The whole subject was introverted. We who were working in it were striving to advance the theory of the calculus of variations as an end in itself, without attention to its relation with other fields of activity.

In contrast, the theory of optimal control attracted great attention as soon as Pontryagin and his followers published it in the late 1950's; and I think that that is as it should be. In my mind, the greatest difference between the Russian approach and ours was in mental attitude. Pontryagin and his students encountered some problems in engineering and in economics that urgently asked for answers. They answered the questions, and in the process they incidentally introduced new and important ideas into the calculus of variations. I think it is excusable that none of us in this room found answers in the 1930's for questions that were not asked until the 1950's. But I for one am regretful that when the questions arose, I did not notice them. Like most mathematicians in the United States, I was not paying attention to the problems of engineers.

In order to discuss this new aspect of the calculus of variations it is convenient to introduce a different method of formulating extremum problems, that includes all the older formulations and also the new problems that arose in the 1950's. A curve $t \to y(t)$ can be regarded as the path of a moving point, and the motion can be controlled by choosing a value of $\dot{y}(t)$ for each t. But with conditioned problems we may not be able to choose $\dot{y}$ arbitrarily. For example, the n components of the vector $\dot{y}(t)$ may have to satisfy some differential equations, fewer than n of them. Also, we may not wish to choose $\dot{y}(t)$ directly, but to fix it by choosing some parameters u that determine $\dot{y}$. Therefore we shall suppose that there exist $n + 1$ functions $(t,y,u) \to f^i(t,y,u)$, defined for all (t,y) in $(n + 1)$-space R^{n+1} and all u in a set $\Omega(t,y)$. We control the curve by choosing a function $t \to u(t)$. If there is an n-vector-valued function $t \to y(t)$ $(a \leq t \leq b)$ such that for all t in $[a,b]$, $u(t) \in \Omega(t,y(t))$ and

$$(2) \quad y(t) = y(a) + \int_a^t f^i(\tau,y(\tau),u(\tau))d\tau \ , \quad (i = 1,\ldots,n) \ ,$$

the function u is an admissible control function and $y(\cdot)$ is the response, or trajectory, corresponding to it. The problem is to find an admissible control $u(\cdot)$ such that the integral

$$\int_a^b f^0(t,y(t),u(t))dt \tag{3}$$

attains its least value among all admissible controls for which the responses satisfy certain end-conditions.

This formulation not only includes all previous ones; it is easier to work with. Magnus Hestenes stated it in 1950; but his results were published in a RAND report with little circulation and at a time when the calculus of variations was at ebb-tide, and they did not attract the attention they deserved. Even earlier, in a paper published in the *Transactions of the American Mathematical Society* in 1933, L. M. Graves had transformed the problem of Lagrange into the control formulation, and had established analogues of the Lagrange multiplier rule (the Euler-Lagrange equation) and the Weierstrass condition. These together are equivalent to the "Pontryagin maximum principle," but only when the set $\Omega(t,y)$ is all of a Euclidean space and the minimizing curve satisfies an annoying condition called "normality." Apparently Pontryagin and his associates did not notice that Graves and Hestenes had both made use of the notation, and they invented it independently. But they introduced one important new feature. They allowed the sets $\Omega(t,y)$ to be closed sets, not demanding that they be open as previous researchers had. To me, as a participant in the older research, it is of interest to distinguish what it was that they adapted from work of their predecessors and what they introduced that was quite new. But I lack the time, and I suspect that most of us lack the interest, for such historical research. However, I do wish to point out that in their book, published in English translation in 1962, Pontryagin and his co-authors made a great step forward in one respect, but a step backward in another. They stated a "maximum principle" that is a generalization of the necessary conditions of Euler and Lagrange, of Legendre and of Weierstrass; but like the mathematicians of the eighteenth century, they gave no sufficient conditions for a minimum, and they stated an existence theorem only for a quite special and simple case. Except for this last, their theory is what L. C. Young calls the "naïve" theory.

Other authors proved existence theorems, but usually under the restrictive condition that the image in R^{n+1} of $\Omega(t,y)$ under the mapping

$$u \to (f^0(t,y,u),\ldots,f^n(t,y,u))$$

is a convex set. Far better results can be obtained by generalizing the problem to allow "relaxed" controls, an obvious extension of the idea of generalized curves. For each t, instead of choosing a point $u(t)$ in $\Omega(t,y(t))$, we choose a probability measure P_t on $\Omega(t,y(t))$. Then equations (2) are replaced by

$$(4)\quad y^i(t) = y^i(a) + \int_a^t \left\{\int_{\Omega(\tau,y(\tau))} f^i(\tau,y(\tau),u)P_\tau(du)\right\}d\tau$$

and the integral to be minimized is

$$(5)\qquad \int_a^b \left\{\int_{\Omega(\tau,y(\tau))} f^0(\tau,y(\tau),u)P_\tau(du)\right\}d\tau \ .$$

Since we shall mention these often, it is expedient to introduce some notation and terminology. We shall sometimes write $P_\bullet$ for the distribution-valued function $t \to P_t$. An *admissible pair* is a pair

$$C = (y(\cdot),P_\bullet) = ((y(t),P_t) : a \leq t \leq b)$$

in which for each t in $[a,b]$, P_t is a probability distribution on $\Omega(t,y(t))$, and the functions $(f^i(t,y(t),u) : u \in \Omega(t,y(t))$ are defined and are integrable with respect to P_t over $\Omega(t,y(t))$, and equations (4) are satisfied. The integral (5) will be denoted by $J(C)$:

$$J(C) = \int_a^b \{\int_{\Omega(t,y(t))} f^0(t,y(t),u)P_t(du)\}dt .$$

For simplicity we shall restrict our attention to the important special case in which $\Omega(t,y)$ is independent of t and y ; we denote it simply by Ω . We shall suppose that F^* denotes a closed set in R^{n+1} and E^* a bounded closed set in R^{2n+2} ; the problem is to find an admissible pair C such that the points $(t,y(t))$ lie in F^* and the end-points $(a,y(a),b,y(b))$ in E^* , and $J(C)$ is the least value of J for all such admissible pairs. If Ω is compact and the f^i are continuous, and some condition is satisfied that guarantees the existence of a minimizing sequence that lies in a bounded subset of F^* , the minimum can be shown to exist. The proof can be found in several research papers and several books, in varying degrees of generality and simplicity.

However, in order to carry out our suggested program of solving the minimizing problem, the necessary conditions for an optimum relaxed control should apply to the kind of optimum that has been shown to exist. This is not taken care of in all books and papers on the subject. However, it was done in 1962 in three independent papers, by J. Warga, by T. Ważewski, and by R. V. Gamkrelidze. It is easily accessible in the books on control theory by L. C. Young and by J. Warga. Young proves an existence theorem somewhat more general than that of the preceding paragraph. For brevity, we shall restrict its generality, but allow its extension to problems in parametric form. A control problem is in parametric form if the control set Ω is a cone, so that whenever u is in Ω so is pu for all nonnegative p ; and the end-conditions are independent of t ; and the functions f^i are independent of t and are positively homogeneous of degree 1 in u , so that

$$f^i(y,pu) = pf^i(y,u)$$

for u in Ω and $p \geqq 0$. It can then be shown that if F^* , E^* and Ω are closed and E^* is bounded, and Ω is independent of y , and the f^i are continuous on $F^* \times \Omega$, and either Ω is bounded or the problem is in parametric form, and there exists a minimizing sequence of admissible pairs for which

$$\int_a^b \{\int_\Omega |u| P_t(du)\} dt$$

is bounded, then there exists an admissible pair $C = (y(\cdot), P_{\bullet})$ for which the integral $J(C)$ has its least value. Moreover, if the problem is in parametric form, the range of t can be taken to be $[0,1]$, and the probability distribution P_t can be so chosen that for a certain positive number L the support of P_t is in $\{u \in \Omega : |u| = L\}$; that is,

$$P_t\{u \in \Omega : |u| \neq L\} = 0 \qquad (0 \leqq t \leqq 1) .$$

For discussing necessary conditions for optimal controls we consider only the case in which for the optimizing pair $(y(\cdot), P_{\bullet})$ the points $(t, y(t))$ lie in the interior of the set F^* . Also, for simplicity we shall consider only the fixed-end-point problem. For these it is proven, in Young's book and elsewhere, that the following theorem holds.

THEOREM Let $((y(t),P_t) : a \leq t \leq b)$ be an optimal pair for the problem described above. For each t in $[a,b]$, let $(t,y(t))$ be interior to F^*, and the support of P_t in a bounded subset of Ω. Then there exist a constant ψ_0, with value 0 or -1, and n absolutely continuous functions $\psi_1,\ldots,\psi_n$ on $[a,b]$, with the following properties.

(i) $\psi_0,\psi_1(t),\ldots,\psi_n(t)$ are not all 0 for any t in $[a,b]$.

(ii) If for each t in $[a,b]$, each y in F^* and each u in Ω we define

$$H(t,y,u) = \psi_0 f^0(t,y,u) + \sum_{i=1}^{n} \psi_i(t) f^i(t,y,u) ,$$

then the maximum value of $H(t,y(t),u)$ on Ω is a continuous function of t on $[a,b]$; and if the problem is in parametric form, the value of this maximum is 0.

(iii) For almost all t in $[a,b]$, the support of P_t is contained in the set on which $H(t,y(t),u)$ attains its maximum value; and by changing P_t on at most a set of measure 0 we can cause this to hold for all t in $[a,b]$.

(iv) The functions ψ_i satisfy

$$d\psi_i(t)/dt = -\int_\Omega \frac{\partial H}{\partial y^i}(t,y(t),u)P_t(du)$$

for almost all t in $[a,b]$.

This theorem and its generalizations are well-known to have interesting applications in problems of optimal control. Many of them are contained in the books by Pontryagin et als., by L. C. Young and by L. D. Berkovitz. But if the optimal control formulation is, as I believe, the modern replacement for the classical calculus of variations, it must be able to provide solutions for the problems of the classical calculus of variations, and it should do so with at most little additional effort. Doubts have been expressed that the ancient problems can be at all conveniently solved by optimal control methods, without transferring back to the old notation. I do not think that that is so, and to bear out my opinion I have worked out several very old problems by optimal control theory. Since the point of the task is to show that a treatment in full detail can be presented without especial difficulty, I lack time and space to present all these calculations in this talk. Instead, I have prepared some sheets [the Appendix] on which I have written out the details of the solutions of four classical problems. Although these problems have been discussed in many books, the treatment there is "naïve," no existence theorems being established. More surprisingly, even from the naïve point of view many of the discussions are incomplete or incorrect. The fourth of the problems is to me much the most interesting. It is the ancient problem of Newton: to find a surface of revolution for which a certain integral, believed by Newton to represent the resistance encountered by that surface when moved through a fluid, has least possible value. Newton did not clearly specify the curves he permitted. So in practically all discussions all piecewise smooth functions $x \to y(x)$ are allowed, and it is then shown that the problem is unsolvable. Only Goursat, in the third volume of his "Cours d'Analyse," points out that it is physically reasonable to assume y monotonic; this might well have been what Newton meant. In no book, not even in Goursat's, is Newton's statement about the solution quoted in full. But the solution of the problem with monotone y exists, and it can be found by the methods of optimal control theory, and it is just what Newton said it is. In this problem it appears that the optimal control procedures are essential; the solution is not accessible by classical procedures.

From what I have just said it is easy to deduce what I believe should be done about the teaching of the calculus of variations. Ordinary undergraduate students of mathematics should be taught a form of control theory simple enough to be understood and general enough to be applicable to many problems. This may call for some quite new chapters in advanced calculus texts, but I think it is not unattainable, and not even very difficult. But another quite different matter is the direction of research. During the twentieth century the interest in what might be called traditional calculus of variations has sunk to a low ebb. I think that that is a natural consequence of the introversion of the subject. Theorems were proved of increasing intricacy, of interest to a steadily shrinking collection of experts in the subject. The newer calculus of variations will go the same way unless its practitioners are sensitive to the questions that arise naturally and demand answers. My own guess is that some of these have to do with the consideration of problems in which random events play an important part. There has, in fact, been a considerable development of stochastic control theory. But both in it and in the deterministic theory I feel that the tendency toward introversion is showing up. The theorems are becoming more baroque. In particular, in quite complicated situations we can show that a solution exists; but only in simple situations can we find a usable approximation to that solution without vast computational effort. The situation is bad in the deterministic case and worse in the stochastic. I am no expert in computation, but I have been told that the direct application of the maximum principle to problems of even moderate complexity is unsatisfactory. This is easy to believe, because the direct application of the maximum principle would require us to find the value of u at which $H(t,y(t),u)$ takes its maximum value, and it is hard to find precisely where a maximum occurs. For stochastic problems the situation is worse. In some cases solutions have been stated which in my opinion are not solutions at all, since they ask the controller to perform infinitely many adjustments of the controls guided by the instantaneous availability of infinitely many bits of information. Others, for example Balakrishnan, have worked on the problem of finding approximate solutions that are humanly attainable. But a great deal remains to be done both in the deterministic and in the stochastic cases. A friend of mine, a logician, once gave a talk in which he proved that no matter how many problems in mathematics have been solved at any given time, there will always remain unanswered questions. We hardly need that theorem. Just within optimal control theory there are *good* problems enough to fill all the time and demand all the brain power that all the available mathematicians can give to them.

APPENDIX: FOUR CLASSICAL PROBLEMS OF THE CALCULUS OF VARIATIONS

The four problems we shall solve are in parametric form. In each, F^* is a closed set in n-space, and y_0 and y_1 are points of n-space, and Ω is a closed cone in r-space. The functions

$$(y,u) \to f^i(y,u) \qquad (i = 1,\ldots,n)$$

are continuous on $F^* \times \Omega$ and are positively homogeneous of degree 1 in u for each fixed y. The minimum of

$$J(C) = \int_a^b \{\int_\Omega f^0(y(t),u)P_t(du)\}dt$$

is sought in the class of all admissible pairs $((y(t),P_t) : a \leqq t \leqq b)$ for which every $y(t)$ is in F^*, and

$$y^i(t) = y_0^i + \int_a^t \{\int_\Omega f^i(y(\tau),u)P_\tau(du)\}d\tau \quad (a \leqq t \leqq b) ,$$

and

$$y^i(b) = y_1^i .$$

For such problems it is known that a minimizing pair $((y(t),P_t) : 0 \leqq t \leqq 1)$ exists, the support of P_t being contained in a set $\{u \in \Omega : |u| = L\}$, provided that there exists a minimizing sequence of admissible pairs $((y_n(t),P_{n,t}) : a_n \leqq t \leqq b_n)$ for which the integrals

$$\int_{a_n}^{b_n} \{\int_\Omega |u|P_{n,t}(du)\}dt$$

are bounded.

It can also be shown, as a corollary of the "maximum principle," that if $((y(t),P_t) : a \leqq t \leqq b)$ is a minimizing pair for this problem, and for each t in $[a,b]$ the support of P_t is bounded and $y(t)$ is interior to F^*, there exist a constant ψ_0 $(= 0$ or $-1)$ and n absolutely continuous functions on $[a,b]$, called $\psi_1,\ldots,\psi_n$, such that $\psi_0,\psi_1(t),\ldots,\psi_n(t)$ never vanish simultaneously and

(i) for t in $[a,b]$ the maximum value of

$$H(t,y,u) = \psi_0 f^0(y,u) + \sum_{i=1}^n \psi_i(t)f^i(y,u) \qquad (u \in \Omega)$$

is 0;

(ii) for almost all t in $[a,b]$, this maximum is attained at each point of the support of P_t ;

(iii) $\psi_i(t) = \psi_i(a) - \int_a^t \{\int_\Omega \frac{\partial H}{\partial y^i}(\tau,y,u)P_\tau(du)\}d\tau$

$(a \leq t \leq b)$.

In all four problems $r = 2$, and to avoid superscripts, we shall write (u,v) instead of (u^1,u^2) . Likewise points in two-space shall be denoted by (x,y) instead of (y^1,y^2) , and points in three-space by (x,y,z) .

I. The Classical Isoperimetric Problem

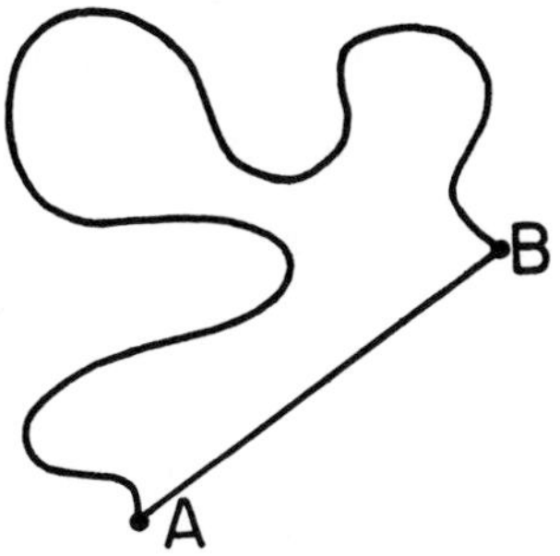

Given a line segment AB , we are to find a curve of length L_0 that goes from B to A and together with AB encloses the greatest possible area.

If the curve is $x = x(t)$, $y = y(t)$, $a \leq t \leq b$, the area enclosed is constant + $\int_a^b [x\dot{y} - y\dot{x}]dt$.

We restate the problem in relaxed-control form. We denote B , A by (x_0,y_0) , (x_1,y_1) , respectively; we define Ω to be the (u,v)-plane; and we define

$$f^0(x,y,z,u,v) = yu - xv ,$$

$$f^1(x,y,z,u,v) = u ,$$

$$f^2(x,y,z,u,v) = v ,$$

$$f^3(x,y,z,u,v) = |(u,v)| = [u^2 + v^2]^{\frac{1}{2}} .$$

The problem is to find an admissible pair $C = ((x(t), y(t), P_t) : a \leq t \leq b)$ with

$$(6) \qquad \begin{aligned} x(a) &= x_0\ , \ y(a) = y_0\ , \ z(a) = 0\ , \\ x(b) &= x_1\ , \ y(b) = y_1\ , \ z(b) = L_0 \end{aligned}$$

for which the integral

$$J(C) = \int_a^b \left\{ \int_\Omega f^0(x(t), y(t), u, v) P_t(du, dv) \right\} dt$$

is minimum. Since the end-points are fixed and the lengths bounded, a solution exists, by the existence theorem on page 22. For it we can assume that for a certain L, the support of P_t is on the circle $|(u,v)| = L$. Since F^* is the whole (x,y)-plane, the necessary conditions on pages 22 and 23 hold; H has the form

$$H(t,x,y,z,u,v) = \psi_0[yu - xv] + \psi_1(t)u + \psi_2(t)v + \psi_3(t)|(u,v)| \ .$$

In particular, H is independent of z, so by equations (iii) ψ_3 is a constant. If this constant were 0 we would have

$$H(t,x,y,z,u,v) = [\psi_0 y + \psi_1(t)]u + [-\psi_0 x + \psi_2(t)]v \ .$$

In order for this linear function to have a maximum on Ω the coefficients of u and v must be 0, so that

$$\psi_1(t) = -\psi_0 y \ , \quad \psi_2(t) = \psi_0 x \ .$$

If $\psi_0 = 0$, all four ψ_i are 0, which is false. If $\psi_0 = -1$, these last equations are incompatible with equations (iii). So $\psi_3 \neq 0$.

We consider two cases.

Case 1 $\psi_0 = 0$.

By equations (iii), ψ_2 and ψ_1 are both constants. Then H assumes its maximum on the circle $|(u,v)| = L$ at just one point, independent of t , so $\dot{x}$ and $\dot{y}$ are constants, and $(x(t),y(t))$ traverses a line-segment from B to A . This is possible, with the end-conditions (1), if and only if L_0 is equal to the distance from B to A . In this case, we have shown that the line-segment BA gives the maximum area. But this is obvious without the discussion, since then there is only one curve that satisfies the conditions for admissibility.

Case 2 $\psi_0 = -1$.

In this case,

$$H(t,x,y,z,u,v) = [-y + \psi_1(t)]u + [x + \psi_2(t)]v + \psi_3(t)|(u,v)| . \tag{7}$$

For (u,v) on the circle $|(u,v)| = L$, this has maximum value at just one point. So for each t , the support of P_t consists of a single point, and the optimal pair is ordinary. By equations (iii) on page 23,

$$\begin{aligned} d\psi_1(t)/dt &= -\int_\Omega [\partial H/\partial x]P_t(du,dv) \\ &= -\int_\Omega vP_t(du,dv) \\ &= -\dot{y}(t) \end{aligned}$$

for almost all t . Likewise, for almost all t ,

$$d\psi_2(t)/dt = \dot{x}(t) .$$

Therefore there exist constants c_1, c_2 such that

$$\psi_1(t) = 2c_2 - y(t) \ , \quad \psi_2(t) = x(t) - 2c_1 \ .$$

We substitute this in (7). By equations (iii), when P_t has only one point in its support, that point is $(\dot{x}(t), \dot{y}(t))$, so H has its maximum there, and its partial derivatives as to u and v are 0. This yields

$$2c_2 - 2y(t) + \psi_3 \dot{x}(t)/L = 0 \ ,$$
$$2x(t) - 2c_1 + \psi_3 \dot{y}(t)/L = 0 \ .$$

Therefore

$$\begin{aligned} \frac{d}{dt}[(x(t) - c_1)^2 + (y(t) - c_2)^2] &= 2(x(t) - c_1)\dot{x}(t) \\ &\quad + 2(y(t) - c_2)\dot{y}(t) \\ &= 0 \ , \end{aligned}$$

and the optimal curve is a circular arc with end-points B and A.

It should be observed that there may be several such arcs. For example, in the figure, BCADEBCA is one, and so is BCA. But an arc that passes more than once through A yields the same area as a curve obtained by rotating through 180° a loop beginning and ending at A, as in the second figure. This new curve BCAD'E'A does not furnish the maximum area, because it is not a circular arc. Therefore neither did the multiply-traversed circular arc from which we

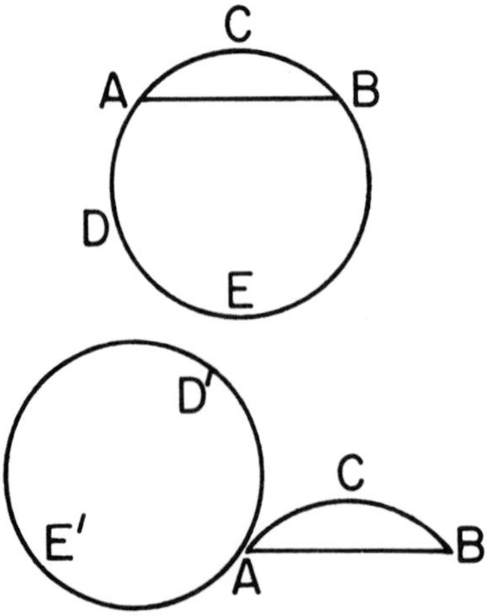

obtained it. The curve that encloses the greatest area is a circular arc without multiple points that goes from B to A .

II. The Brachistochrone

We next consider the problem of the brachistochrone, first proposed by John Bernouilli in 1696.

Let us choose axes in the plane with the y-axis vertically downward. A bead descends by gravity, starting from rest, along a frictionless wire beginning at $(0,0)$ and ending at a point (x_1,y_1) with $x_1 > 0$ and $y_1 \geqq 0$. It is required to find the shape of the wire along which the time of descent will be shortest.

In order for the wire to be traversible at all it must lie in the half-plane

$$F^* = \{(x,y) : y \geqq 0\} .$$

Along a curve $x = x(t)$, $y = y(t)$ $(a \leqq t \leqq b)$ in F^* with absolutely continuous $x(\cdot)$ and $y(\cdot)$ the time of descent is proportional to

$$\int_a^b \left[\frac{\dot{x}(t)^2 + \dot{y}(t)^2}{y(t)} \right]^{\frac{1}{2}} dt ;$$

we are to minimize this. We extend the problem to relaxed-control form. We define F^* as above, and define Ω to be the (u,v)-plane. Admissible pairs are those pairs $C = ((x(t),y(t),P_t) : a \leqq t \leqq b)$ that satisfy

$$x(t) = \int_a^t \{\int_\Omega uP_\tau(du,dv)\}d\tau ,$$

$$y(t) = \int_a^t \{\int_\Omega vP_\tau(du,dv)\}d\tau$$

and the end-conditions

$$x(b) = x_1 \ , \quad y(b) = y_1 \ ,$$

and have $(x(t),y(t))$ in F^* for all t. Among these we seek a pair C for which the integral

$$J(C) = \int_a^b \{\int_\Omega f^0(x(t),y(t),u,v)P_t(du,dv)\}dt$$

is a minimum, where

$$f^0(x,y,u,v) = [u^2 + v^2]^{\frac{1}{2}}/y^{\frac{1}{2}} \ .$$

Although the existence theorem on page 22 does not apply to this problem, the integrand being discontinuous at $y = 0$, a slight modification using a limit process can be used to show that a minimizing pair exists, with $a = 0$, $b = 1$, and the support of P_t contained in the set $\{|(u,v)| = L\}$. We omit the details of this proof.

Suppose that there is a t^* in the open interval $(0,1)$ for which $y(t^*) = 0$. Since $J(C)$ is finite, y is not identically 0 on $[0,t^*]$ or on $[t^*,b]$. We can therefore find a positive number c and numbers t_1,t_2x such that $0 < t_1 < t^* < t_2 < b$, and $y(t_1) = y(t_2) = c$, and $y(t) < c$ on (t_1,t_2). Then

$$\int_{t_1}^{t_2} \{\int_\Omega y^{-\frac{1}{2}}[u^2 + v^2]^{\frac{1}{2}}P_t(du,dv)\}dt > \int_{t_1}^{t_2} \{\int_\Omega c^{-\frac{1}{2}}uP_t(du,dv)\}dt$$

$$= \int_{t_1}^{t_2} c^{-\frac{1}{2}}\dot{x}(t)dt \ .$$

So if we replace C by a pair C^* in which on the interval $[t_1,t_2]$ the relaxed pair $(x(\cdot),y(\cdot),P_\cdot)$ is replaced by the ordinary curve $x = x(t)$, $y = c$, C^* is an admissible pair, and $J(C^*) < J(C)$, which is impossible. So $y(t) > 0$ for $0 < t < 1$.

For every subinterval $[a,b]$ of the open interval $(0,1)$, the admissible pair $((x(t),y(t),P_t) : a \leq t \leq b)$ minimizes the integral

$$\int_a^b \{\int_\Omega y(t)^{-\frac{1}{2}}[u^2 + v^2]^{\frac{1}{2}} P_t(du,dv)\}dt$$

in the class of admissible pairs with the same end-points, and the points of its trajectory lie in the interior of F^*. Therefore the necessary conditions on pages 22 and 23 are satisfied. H has the form

$$H(t,x,y,u,v) = \psi_0 y^{-\frac{1}{2}}[u^2 + v^2]^{\frac{1}{2}} + \psi_1(t)u + \psi_2(t)v ,$$

and it attains its maximum at each point in the support of P_t. If ψ_0 were 0, H could have no maximum, so $\psi_0 = -1$. This implies that H can have its maximum value at only one point of the circle $|(u,v)| = L$, so the curve is ordinary, and the one point in the support of P_t is (by equations (iii)) the point $(\dot{x}(t),\dot{y}(t))$ for almost all t.

By the first of equations (iii), $\psi_1(t)$ is a constant. If it were 0, the maximum value of H on $|(u,v)| = L$ would occur at $(0,L)$ or at $(0,-L)$, yielding $\dot{x}(t) = 0$ for almost all t. But then $x(b) = x(a)$, and by letting a tend to 0 and b to 1 we find $x(1) = x(0)$, which is false. So $\psi_1 \neq 0$.

For almost all t, $(\dot{x}(t),\dot{y}(t))$ is the only point in the support of P_t, so $H(t,x(t),y(t),u,v)$ attains its maximum at that point. Therefore at $(\dot{x}(t),\dot{y}(t))$

$$0 = \partial H/\partial u$$
$$= -y(t)^{-\frac{1}{2}}\dot{x}(t)[\dot{x}(t)^2 + \dot{y}(t)^2]^{-\frac{1}{2}} + \psi_1 .$$

Since $\psi_1 \neq 0$, this implies that $\dot{x}(t)$ is bounded away from 0 and has the same sign as ψ_1 (necessarily positive). So the function $t \to x(t)$ $(0 \leq t \leq 1)$ has an inverse $x \to t(x)$ $(0 \leq x \leq x_1)$, and the optimizing curve has the representation

$$x \to Y(x) = y(t(x)) \qquad (0 \leq x \leq x_1) .$$

By the preceding equation,

$$Y(x)[1 + (dY/dx)^2] = \psi_1^{-2} .$$

This is a familiar equation. Its solution is a cycloid; see, for example, Elsgolc, page 38.

III. The Surface of Revolution of Least Area

Euler proposed the problem of finding the curve that joins two points in the (x,y)-plane and, among such curves, generates when revolved about the x-axis that surface that has least area. If these end-points are (x_0,y_0) and (x_1,y_1), and the curve has a representation $x = x(t)$, $y = y(t)$, $(a \leq t \leq b)$ in which $x(\cdot)$ and $y(\cdot)$ are absolutely continuous, the area of the surface of revolution is

$$(8) \qquad 2\pi \int_a^b |y(t)|[\dot{x}(t)^2 + \dot{y}(t)^2]^{\frac{1}{2}}dt ;$$

and this is the integral to be minimized. It is rather obvious that this is the same problem as minimizing the integral

$$(9) \qquad \int_a^b y(t)[\dot{x}(t)^2 + \dot{y}(t)^2]^{\frac{1}{2}}dt$$

in the class of absolutely continuous functions x,y with the given end-points and lying in the half-plane

(10) $$F^* = \{(x,y) : y \geqq 0\} ;$$

see, for example, Bliss' Carus monograph "Calculus of Variations," p. 89. This apparently minor point has led to false assertions in several books. For example, both in Gelfand and Fomin and in Elsgolc we find the statement that the area is given by (3) with the absolute value sign omitted; this is incorrect, and without the qualification $y \geqq 0$ (which they do not mention) the integral (9) has no lower bound. Bliss correctly states the problem, but on p. 90 says that the necessary conditions previously deduced apply without change to this problem, which is incorrect because the minimizing curve can lie in part along the boundary of F^*.

We restate the problem as a relaxed-control problem. Let F^* be defined by (10), and let Ω be the (u,v)-plane. Define

$$f^0(x,y,u,v) = y[u^2 + v^2]^{\frac{1}{2}} \quad ((x,y) \text{ in } F^*, \text{ all } (u,v)).$$

Among all admissible pairs $((x(t),y(t),P_t) : a \leqq t \leqq b)$ that have $(x(t),y(t))$ in F^* for all t, and have end-points $x(a) = x_0$, $y(a) = y_0$, $x(b) = x_1$, $y(b) = y_1$, and satisfy equations

(11) $$x(t) = x_0 + \int_a^t \{\int_\Omega uP_\tau(du,dv)\}d\tau$$
$$y(t) = y_0 + \int_a^t \{\int_\Omega vP_\tau(du,dv)\}d\tau ,$$

to find one that minimizes the integral

$$J(C) = \int_a^b \{\int_\Omega f^0(x(t),y(t),u,v)P_t(du,dv)\}dt .$$

Let m be the infimum of $J(C)$ in the class of pairs admitted. A minimizing sequence is a sequence of admissible pairs

$$C_n = ((x_n(t), y_n(t), P_{n,t}) : a_n \leqq t \leqq b_n) \quad (n = 1,2,3,\dots)$$

satisfying the requirements and such that the integrals $J(C_n)$ tend to m. We distinguish two cases.

Case 1 There exists a minimizing sequence for which the minimum value of $y_n(t)$ on $[a_n, b_n]$ is arbitrarily near 0.

In this case we can choose a subsequence (which without loss of generality we may take to be the original sequence) for which

$$(12) \qquad \lim_{n\to\infty} \min\{y_n(t) : a_n \leqq t \leqq b_n\} = 0 .$$

Let c_n be a point in $[a_n, b_n]$ at which y_n attains its least value. Then

$$\int_{a_n}^{c_n} \{\int_\Omega y_n(t)[u^2 + v^2]^{\frac{1}{2}} P_{n,t}(du,dv)\}dt$$

$$\geqq \int_{a_n}^{c_n} \{\int_\Omega y_n(t)[-u] P_{n,t}(du,dv)\}dt$$

$$= - \int_{a_n}^{c_n} y_n(t)\dot{y}_n(t)dt$$

$$= [y_0^2 - y_n(c_n)^2]/2 \qquad ,$$

and similarly

$$\int_{c_n}^{b_n} \{\iint_\Omega y_n(t)[u^2 + v^2]^{\frac{1}{2}} P_{n,t}(du,dv)\}dt$$

$$\geqq \int_{c_n}^{b_n} \{\iint_\Omega y_n(t) u P_{n,t}(du,dv)\}dt$$

$$= \int_{c_n}^{b_n} y_n(t)\dot{y}_n(t)dt$$

$$= [y_1^2 - y_n(c_n)^2]/2 \quad .$$

By adding these we obtain

$$[y_0^2 + y_1^2 - 2y_n(c_n)^2]/2 \leqq J(C_n) \ .$$

Since $y_n(c_n)$ is the least value of y_n and $J(C_n)$ tends to m, by (12)

$$[y_0^2 + y_1^2]/2 \leqq m \ .$$

The left member of this inequality is $J(C)$ for the ordinary curve C which is the polygon with successive vertices (x_0,y_0), $(x_0,0)$, $(x_1,0)$, (x_1,y_1), so that polygon minimizes $J(C)$ in the class of admissible pairs with the given end-points.

Case 2 For each minimizing sequence, all y_n have a common positive lower bound.

Choose a minimizing sequence, with the same notation as before. Let $c > 0$ be a lower bound for all the y_n. Then

$$\int_{a_n}^{b_n} \{\int_\Omega [u^2 + v^2]^{\frac{1}{2}} P_{n,t}(du,dv)\}dt$$

$$\leq c^{-1} \int_{a_n}^{b_n} \{\int_\Omega y_n(t)[u^2 + v^2]^{\frac{1}{2}} P_{n,t}(du,dv)\}dt$$

$$= J(C_n)/c$$

So the integrals are bounded, and by the existence theorem on page 22 a minimizing admissible pair exists, and it can be chosen to be a pair for which $a = 0$ and $b = 1$, and the support of P_t is contained in a circle $|(u,v)| = L$. For this pair, the minimum of y is positive; otherwise, a sequence of infinitely many repetitions of C would be a minimizing sequence in which the minima of the y_n are arbitrarily near 0, and in the case we are considering that cannot happen. Since $y(t)$ is always positive, the trajectory lies in the interior of F^*, and the necessary conditions on pages 22 and 23 must be satisfied. Let ψ_0 (= 0 or -1), $\psi_1(t)$, $\psi_2(t)$ be the multipliers, so that

$$H(t,x,y,u,v) = \psi_0 y(t)[u^2 + v^2]^{\frac{1}{2}} + \psi_1(t)u + \psi_2(t)v .$$

If ψ_0 were 0 this would be a linear function and have no maximum on Ω. Therefore $\psi_0 = -1$.

On the circle $\{|(u,v)| = L\}$ the function $H(t,x(t),y(t),u,v)$ attains its maximum at only one point, and by equations (11) this is $(\dot{x}(t),\dot{y}(t))$ for almost all t. So the optimal control is ordinary. Since H is independent of x, by the first of equations (iii) ψ_1 is constant. If this constant were 0, the maximum of H would occur at either $(0,L)$ or at $(0,-L)$, and in either case $\dot{x}(t) = 0$. This is impossible, since it would imply $x(1) = x(0)$, which is false. So

$\psi_1 \neq 0$. The maximum value of H occurs at $(\dot{x}(t),\dot{y}(t))$, so its partial derivative with respect to u vanishes at that point:

$$0 = -y(t)[\dot{x}^2(t) + \dot{y}^2(t)]^{-\frac{1}{2}}\dot{x}(t) + \psi_1 .$$

The quantity in square brackets has value L^2 , so $\dot{x}(t)$ is bounded away from 0 , and y can be expressed as a function of x . The preceding equation then yields

$$y(t) = \psi_1[1 + (dy/dx)^2]^{\frac{1}{2}} .$$

The solution of this is well known to be

$$y = \psi_1 \cosh[(x - b)/\psi_1] ,$$

where b is a constant. This is the equation of a catenary. It is shown in many places that there are at most two such catenaries through (x_0,y_0) and (x_1,y_1) . So the absolute minimum of the surface of revolution is given by the polygon or by one of these at most two catenaries.

IV. The Solid of Revolution of Least Resistance

In his Principia, Isaac Newton discussed the resistance encountered by bodies moving through a fluid. Although he never stated a law of resistance specifically, his reasoning in Book II, Proposition XXXIV, Theorem XXVIII (in the discussion of the resistance of a sphere) leads to the following law:

The resisting pressure at any point of the surface *not sheltered from the fluid by some part of the body* is proportional to the square of the component of the velocity along the normal to the surface.

This is Forsythe's formulation, except for the italicized words, which I have added. They are justified by Newton's reasoning, and by his ignoring pressure on the sheltered hemisphere. If a solid of revolution is generated by revolving a curve $x = x(t)$, $y = y(t)$ about the y-axis and moves in the direction of the positive y-axis, and x is not monotonic increasing, there will be sheltered arcs such as ABC . By Newton's reasoning, there will be no resisting pressure on this part of the solid, and we can replace the arc ABC by the line segment AC without affecting the resistance. So we may, and henceforth shall assume that $x(\cdot)$ is non-decreasing.

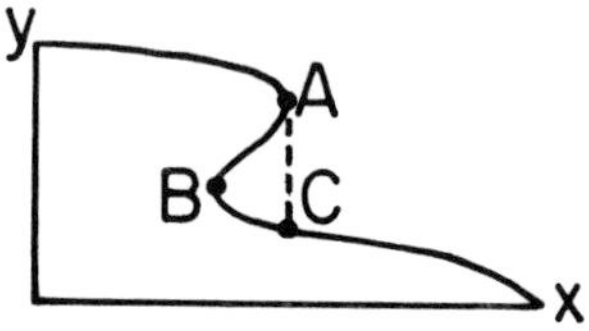

Newton's assertion about the solid of least resistance is contained in the following two paragraphs of the Scholium following the theorem just cited.

"Incidentally, ..., it follows from the above that, if the solid ADBE be generated by the convolution of an elliptical or oval figure ADBE about its axis AB , and the generating figure be touched by three right lines FG , GH , HI in the points F , B , and I , so that GH shall be perpendicular to the axis in the point of contact B , and FG , HI may be inclined to GH in the angles FGB , BHI of 135 degrees: the solid arising from the convolution of the figure ADFGHIE about the same axis AB will be less resisted than the former solid, provided that both move forwards in the direction of their axis AB , and that the extremity B of each go forward. This

Proposition I conceive may be of use in the building of ships.

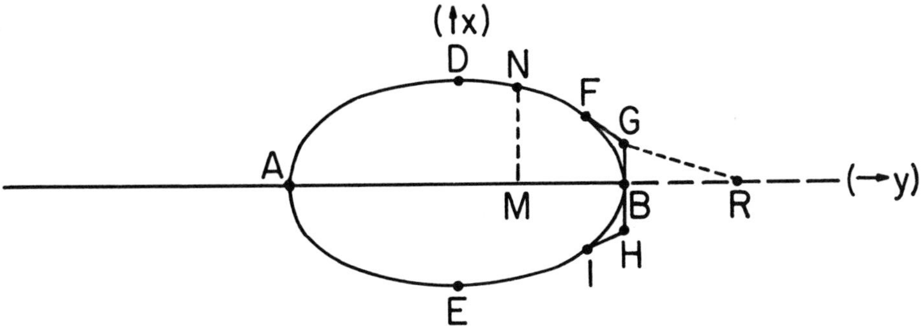

"If the figure DNFG be such a curve, that if, from any point thereof, as N , the perpendicular NM be let fall on the axis AB , and from the given point G there be drawn the right line GR parallel to a right line touching the figure in N , and cutting the axis produced in R , MN becomes to GR as GR^3 to $4BR \cdot GB^2$, the solid described by the revolution of this figure about its axis AB , moving in the before-mentioned rare medium from A toward B , will be less resisted than any other circular solid whatsoever, described of the same length and breadth."

Let us equip Newton's figure with an x-axis vertically upward through D and a y-axis along AR , positive in that direction. Then B is $(0,y_0)$ and D is $(x_1,0)$. We denote the length of BG by the perhaps startling symbol $2\psi_2$. The statements in the quoted paragraphs transform into twentieth century notation thus. Let the optimal curve BGFND joining B to D be the graph $x \to y(x)$ $(0 \leqq x \leqq x_1)$. Then $y(x)$ is constantly y_0 for $0 \leqq x \leqq 2\psi_2$; at $2\psi_2$, the right derivative of y is -1 ; and on $[2\psi_2,x_1]$ the function $y(\cdot)$ satisfies the equation

$$x = -\psi_2[1 + (dy/dx)^2]^2/2(dy/dx) . \tag{13}$$

In none of the books on the Calculus of Variations that I have consulted have I found any such statement. They all deduce equation (7), but omit all reference to the line-segment BG and the corner at G .

Newton did not specify the class of comparison curves allowed. We shall allow only curves $x = x(t)$, $y = y(t)$ in which x is non-decreasing (which, as we have seen, is very nearly implied by Newton's discussion) and $y(\cdot)$ is non-increasing. Newton might have been willing to accept this reasoning. If y is not monotonic non-increasing, the solid of revolution will have a trough generated by the revolution of an arc such as DEF . When the body moves in the direction of the positive y-axis, this trough will fill with stagnant fluid, and the line-segment DF will be the effective surface, giving a non-increasing y . Newton makes no such statement. But in the first paragraph quoted above, he specifically considers "oval" solids, and in the Principia and in a letter (presumably written to David Gregory in 1694) only convex figures appear. So we feel that we are not misrepresenting Newton when we attach his name to this problem:

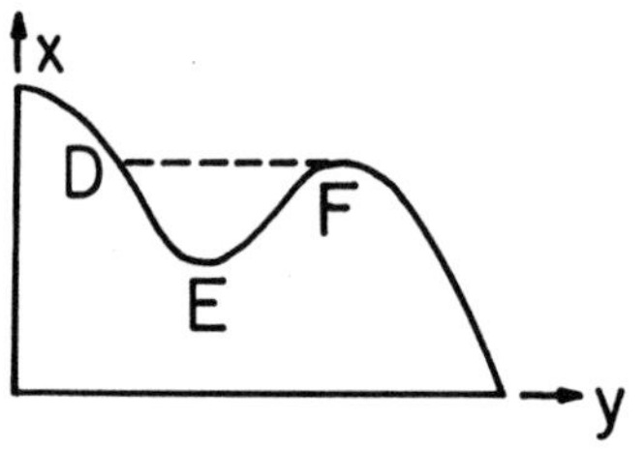

To find a curve $x = x(t)$, $y = y(t)$, $a \leqq t \leqq b$ in which $x(\cdot)$ is non-decreasing and $y(\cdot)$ is non-increasing, and

$$x(a) = 0 \ , \ y(a) = y_0 \ , \ x(b) = x_1 \ , \ y(b) = y_1 \qquad (x_1 \text{ and } y_0 \text{ positive}) \tag{14}$$

and which in that class of curves generates the surface of

revolution about the y-axis that offers least resistance to motion in the direction of the positive y-axis, the law of resistance being that stated above.

We shall show that this problem has a solution, and that the solution is exactly what Newton asserted it to be.

According to Newton's law of resistance, the resistance at a given velocity is proportional to

$$J(C) = \int_a^b \frac{x(t)\dot{x}(t)^3}{\dot{x}(t)^2 + \dot{y}(t)^2}\, dt \,, \tag{15}$$

and this is the integral to be minimized. We restate the problem in relaxed control form.

Let Ω be the control region

$$\Omega = \{(u,v) : u \geqq 0 \,,\ v \geqq 0\} \,.$$

Define

$$f^0(x,y,u,v) = xu^3/[u^2 + v^2] \,,$$

$$f^1(x,y,u,v) = u \,,$$

$$f^2(x,y,u,v) = -v \,.$$

A pair $((x(t),y(t),P_t) : a \leqq t \leqq b)$ is admissible if P_t is a probability distribution on Ω , and

$$x(t) = \int_a^t \{\int_\Omega f^1(x(t),y(t),u,v)P_\tau(du,dv)\}d\tau \,,$$

$$y(t) = y_0 + \int_a^t \{\int_\Omega f^2(x(t),y(t),u,v)P_\tau(du,dv)\}d\tau$$

$$(a \leqq t \leqq b) \,,$$

and the end-conditions (14) are satisfied. Among all admissible pairs we seek one that minimizes

$$J(C) = \int_a^b \{\int_\Omega f^0(x(t),y(t),u,v)P_t(du,dv)\}dt .$$

The simpler existence theorems do not apply to this problem, nor to the analogous problem in ordinary controls, because f^0 is not a convex function. But by the existence theorem on page 22 an admissible (relaxed) pair does exist that minimizes $J(C)$ among all such pairs. It has also the property that $a = 0$, and $b = 1$, and for all t, the support of P_t is contained in the circular arc

$$\Omega_L = \{(u,v) : 0 \leq u , 0 \leq v , u^2 + v^2 = L^2\} .$$

Since F^* in this problem is the whole plane, the necessary conditions on pages 22 and 23 are satisfied. As before, we define

$$H(t,x,y,u,v) = \psi_0 f^0(x,y,u,v) + \sum_{i=1}^{2} \psi_i(t)f^i(x,y,u,v) ;$$

the ψ_0 and ψ_i have the properties on pages 22 and 23. In particular, since H is independent of y, ψ_2 is a constant.

We shall first prove that $\psi_0 \neq 0$. Suppose it were. Then

$$H = \psi_1(t)u - \psi_2 v .$$

If H were positive at some (u,v) in Ω, it would be unbounded on the set of points (pu,pv) with $p > 0$, which is impossible because H has a maximum attained on Ω_L. So H is nonpositive, and its maximum value on Ω is 0, attained at $(0,0)$. So H is nonpositive at $(1,0)$ and at $(0,1)$, and

$$\psi_1(t) \leq 0 \leq \psi_2 .$$

If $\psi_2 > 0$, the maximum value of H on Ω_L occurs at $(L,0)$ only, so only $(L,0)$ is in the support of P_t. Therefore

$\dot{y}(t) = 0$ for almost all t, and $y(t) = y_0$ for all t, which is false. If $\psi_2 = 0$, $\psi_1(t)$ must never be 0, since ψ_0, ψ_1 and ψ_2 never vanish simultaneously. So $\psi_1(t) < 0$ for all t. Then the only point of Ω_L at which H is maximum is $(0,L)$. This implies $x(t) = 0$ for all t, which is false. So the assumption $\psi_0 = 0$ leads to contradiction, and ψ_0 is -1. Now

(16) $$H(t,x,y,u,v) = -xu^3/[u^2 + v^2] + \psi_1(t)u - \psi_2 v .$$

We next prove $\psi_2 \neq 0$. If $\psi_2 = 0$, let t be any point at which $x(t) > 0$, and let (u^*,v^*) be in the support of P_t. Then H has its maximum value 0 at (u^*,v^*), so

$$-x(t)u^{*3}/[u^{*2} + v^{*2}] + \psi_1(t)u^* = 0 .$$

If $u^* > 0$, this would imply

$$H(t,x(t),y(t),u^*,v^* + 1) > 0 ,$$

which is impossible. So $u^* = 0$. Therefore the only point in the support of P_t is $(0,L)$, and $dx/dt = 0$ at almost all points at which $x(t) > 0$. This is incompatible with the end-conditions (14), so the assumption $\psi_2 = 0$ has led to contradiction. Therefore $\psi_2 > 0$. This, in turn, implies $\psi_1(0) = 0$. For if not, then $\psi_1(0) < 0$, and the function

$$H(0,x(0),y(0),u,v) = \psi_1(0)u - \psi_2 v$$

would assume its maximum value 0 only at $(0,0)$, not at any point in Ω_L. It also implies that $(0,L)$ is not in the support of P_t for any t in $[0,1]$; for

$$H(t,x(t),y(t),0,L) = -\psi_2 L < 0 .$$

Then

$$\dot{x}(t) = \int_\Omega uP_t(du,dv) > 0$$

for almost all t, so $x(t)$ is strictly increasing.

For $t > 0$, no point (u,v) of Ω with $0 < v < u$ is in the support of P_t. For suppose $0 < v < u$. Define $\theta = v/u$. Then

$$\begin{aligned} 0 &= H(t,x(t),y(t),u,v) \\ &= -x(t)u^3/[u^2 + v^2] + \psi_1(t)u - \psi_2 v \\ &= -x(t)u/[1 + \theta^2] + \psi_1(t)u - \psi_2\theta u , \end{aligned}$$

$$H(t,x(t),y(t),u,0) = -x(t)u + \psi_1(t)u ,$$

$$H(t,x(t),y(t),u,u) = -x(t)u/2 + \psi_1(t)u - \psi_2 u .$$

From the last two equations,

$$\begin{aligned} &(1 - \theta)H(t,x(t),y(t),u,0) + \theta H(t,x(t),y(t),u,u) \\ &= -x(t)(1 - \theta/2)u + \psi_1(t)u - \psi_2\theta u . \end{aligned}$$

Since

$$(1 - \theta/2) - (1 + \theta^2)^{-1} = -\theta(1 - \theta)^2/2(1 + \theta)^2 < 0 ,$$

this implies that one of the numbers $H(t,x(t),y(t),u,u)$, $H(t,x(t),y(t),u,0)$ is positive, in contradiction to the fact that the maximum value of H is 0. So no point (u,v) with $0 < v < u$ is in the support of P_t for any t in $[0,1]$.

Let A be the set of t in $[0,1]$ such that $(L,0)$ is in the support of P_t. If t is in A,

$$\begin{aligned} 0 &= H(t,x(t),y(t),L,0) \\ &\geqq H(t,x(t),y(t),L,L) , \end{aligned}$$

so

$$-x(t)L + \psi_1(t)L - 0 \geqq -x(t)(L/2) + \psi_1(t)L - \psi_2 L .$$

This implies

$$x(t) \leqq 2\psi_2 .$$

If t is in $[0,1]\backslash A$, the support of P_t contains some point (u,v) with $v \neq 0$, therefore with $0 < u \leqq v$. For all t we have by equations (iii) on page 23

$$\begin{aligned}\psi_1(t) &= \int_0^t \{\int_\Omega \frac{u^3}{u^2 + v^2} P_\tau(du,dv)\}d\tau \\ &\leqq \int_0^t \{\int_\Omega uP_\tau(du,dv)\}d\tau \\ &= x(t) .\end{aligned}$$

So if the support of P_t contains (u,v) with $v > 0$,

$$\begin{aligned}0 &= H(t,x(t),y(t),u,v) \\ &= -x(t)u^3/[u^2 + v^2] + \psi_1(t)u - \psi_2 v \\ &\leqq x(t)\{-u^3/[u^2 + v^2] + u\} - \psi_2 v \\ &= \{x(t)uv/[u^2 + v^2] - \psi_2\}v .\end{aligned}$$

Hence

$$\psi_2 \leqq x(t)\{uv/[u^2 + v^2]\} \leqq x(t)\{\tfrac{1}{2}\} ,$$

and therefore

$$x(t) \geqq 2\psi_2 .$$

So all points t with $x(t)$ in $[0,2\psi_2]$ are in A , and all points t with $x(t)$ in $(2\psi_2,1]$ are in $[0,1]\backslash A$. For t with $x(t)$ in $[0,2\psi_2)$ the support of P_t consists of $(L,0)$ alone, so

$$x(t) = \int_0^t \{\int_\Omega uP_\tau(du,dv)\}d\tau$$

$$= \int_0^t Ld\tau = Lt \qquad ,$$

$$y(t) = y_0 - \int_0^t \{\int_\Omega vP_\tau(du,dv)\}d\tau$$

$$= y_0 \qquad .$$

So if we define $t^* = 2\psi_2/L$

$$x(t^*) = 2\psi_2 ,$$

and $y(t)$ is constantly y_0 on $[0,t^*]$. For $t < t^*$, $(L,0)$ is in the support of P_t , so

$$H(t,x(t),y(t),L,0) = 0 .$$

By continuity,

$$H(t^*,x(t^*),y(t^*),L,0) = 0 .$$

For $t > t^*$, the support of P_t is contained in $\{(u,v)$ in $\Omega : v \geqq u > 0\}$. But when $v \geqq u > 0$,

$$\partial^2H/\partial v^2 = 2x(t)u^3(u^2 + v^2)^{-3}(u^2 - 3v^2) < 0 .$$

If the maximum value of H were attained at two different points of Ω_L , there would be two rays $v = m_1u$, $v = m_2u$ $(m_2 > m_1 \geqq 1)$ on which H and its first partial derivatives vanish. That is,

$$\frac{\partial}{\partial v} H(t,x(t),y(t),1,m_1) = \frac{\partial}{\partial v} H(t,x(t),y(t),1,m_2) = 0 .$$

But this is impossible, since

$$\frac{\partial^2 H}{\partial v^2}(t,x(t),y(t),1,v) < 0 \qquad (m_1 \leqq v \leqq m_2) .$$

Therefore the support of P_t consists of a single point of the arc Ω_L . The optimal pair is an ordinary pair.

We have shown that

$$x(t^*) = \psi_1(t^*) = 2\psi_2 ,$$

so

$$H(t^*,x(t^*),y(t^*),u,v) = -2\psi_2 u^3/[u^2 + v^2] + 2\psi_2 u - \psi_2 v .$$

Therefore

$$(17)\qquad \begin{aligned} &H(t^*,x(t^*),y(t^*),L,0) = 0 , \\ &H(t^*,x(t^*),y(t^*),L/\sqrt{2},L/\sqrt{2}) = 0 . \end{aligned}$$

For each point t in $(t^*,1]$, the support of P_t consists of a single point of Ω_L , which we denote by $(u(t),v(t))$. Let $(\alpha(t),\beta(t))$ be any point that is the limit of $(u(t_n),v(t_n))$ for some sequence of points $t_1,t_2,t_3,\ldots$ of $(t^*,1]$ tending to t . Then $\beta(t) \geqq \alpha(t)$. By continuity

$$\begin{aligned} H(t,x(t),y(t),\alpha(t),\beta(t)) &= \lim_{n\to\infty} H(t_n,x(t_n),y(t_n),u(t_n),v(t_n)) \\ &= 0 . \end{aligned}$$

But there is only one point $(\alpha(t),\beta(t))$ on Ω_L with $\beta(t) \geqq \alpha(t)$ that satisfies this. If $t > t^*$, $(\alpha(t),\beta(t))$ has to be $(u(t),v(t))$. If $t = t^*$, by (17)

$$\alpha(t) = L/\sqrt{2} , \beta(t) = L/\sqrt{2} .$$

So the limit of $(u(t),v(t))$ as t tends to t^* from above is $(L/\sqrt{2},L/\sqrt{2})$.

For t in $(t^*,1]$, the support of P_t contains a single point, which for almost all t has to be $(\dot{x}(t),\dot{y}(t))$. Since H has its maximum value at this point, its partial derivatives vanish there, so

(18) $$2x(t)\dot{x}(t)^3\dot{y}(t)[\dot{x}(t)^2 + \dot{y}(t)^2]^{-2} + \psi_2 = 0 .$$

This implies that $\dot{x}$ cannot be 0 , so y can be written as a function of x , and equation (18) amended accordingly. We now have accumulated the following information about the minimizing function, written as $x \to y(x)$:

For $0 \leqq x \leqq 2\psi_2$, $y(x) = y_0$. At $x = 2\psi_2$, there is a corner: the right derivative of y with respect to x is -1 . For $x > 2\psi_2$,

(19) $$x = -\psi_2[1 + (dy/dx)^2]^2/2[dy/dx] .$$

This is exactly what Newton stated the solution to be.

Parametric equations for the part of the curve to the right of $x = 2\psi_2$ are easily obtained. If we define $\tau = v(t)/u(t)$, by (18) or (19) we find that for almost all t the function $X(\tau) = x(t(\tau))$ satisfies

(20) $$X(\tau) = (1 + \tau^2)^2\psi_2/2\tau .$$

From this,

$$dX/dt = (\psi_2/2)[-\tau^{-2} + 2\tau + 3\tau^2] .$$

Let $Y(\tau) = y(t(\tau))$. Since $dY/d\tau = [v/u]dX/d\tau = \tau[dX/d\tau]$,

$$dY/d\tau = (\psi_2/2)[-\tau^{-1} + 2\tau + 3\tau^3] ,$$

whence

(21) $$Y(\tau) = C + (\psi_2/2)[-\log \tau + \tau^2 + 3\tau^4/4] .$$

We let τ approach 1 ; then $X(\tau)$ and $Y(\tau)$ approach $x(t^*) = 2\psi_2$ and $y(t^*) = y_0$, respectively, and from (21) we obtain

$$C = y_0 - (\psi_2/2)(7/4) .$$

Substituting this in (21) yields

$$Y(\tau) = y_0 + (\psi_2/2)[-\log\tau + \tau^2 + 3\tau^4/4 - 7/4] .$$

This and (20) are parametric equations for the part of the curve to the right of the straight section $y = y_0$ $(0 \leqq x \leqq 2\psi_2)$.

ADDED IN PROOF Professor John Burns has pointed out to me that Newton's least-resistance problem is discussed in "Applied Optimal Control" by A. E. Bryson and Y.-C. Ho (pp. 52-55). Their treatment is of the type we have called "naïve," and also it is not rigorous. But they arrive at Newton's solution, as given above, and they exhibit graphs for three special cases. Also they assert that Newton's formula for the resistance, while inaccurate for subsonic speeds, is very good at speeds above the speed of sound.

CONTROL THEORY, VARIATIONAL PRINCIPLES, AND OPTIMAL DESIGN OF ELASTIC SYSTEMS

Vadim Komkov

Texas Tech University
Lubbock, Texas

0. INTRODUCTORY COMMENTS

The control theory for systems with distributed parameters has been considerably advanced in recent years. The particular application to the control of vibrations of elastic systems has been developed around the original ideas of Pontryagin et al ([1], [2], [3], [4], [5]), by Russell ([6], [7]), Komkov ([8], [9], [10], [11], [12], [13]), Barnes ([14]). No consistent theory has been developed for engineering optimization problems, despite the fact that many similarities can be observed both in formulation and mathematical modelling of the problems, and in the form of equations derived as necessary conditions for the existence of an optimal design.

Some partial results have been obtained by applying modern analytical methods, and in some cases the problems of design

ISBN 0-12-632250-3

optimization have been reduced to a form whereby a direct application of Pontryagin's principle is possible. Most papers found in engineering literature concern themselves with approximate numerical procedures leading to optimization of design, and only scattered attempts exist in the literature towards a unified mathematical theory of optimization of design.

1. AN ENGINEERING DESIGN OPTIMIZATION PROBLEM

We offer here a specific example. The problem is well-known and has been considered by some outstanding engineers and mathematicians in the past. The problem of a slender column design goes back to Leonard Euler. The "best column" shape has been considered by von Karman, and more recently by J. B. Keller and F. I. Niordson [29].

In an Army Handbook [17] E. J. Haug has made an observation that some problems of design optimization can be restated in the language of control theory, and that the theoretical development of classical control theory can be applied to some problems concerning the optimization of design. One example given in [7] is the optimization of the shape of a column. It is unfortunate that his approach and numerous examples are given in an obscure publication.

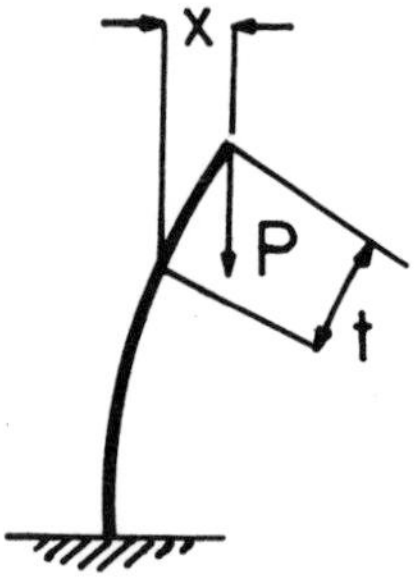

Figure 1: Buckling of a column.

A geometry of the cross-sectional area is determined so that the cross-section $A(t)$ depends on a single design parameter $u(t)$. We design a cross-section of a column of length T, wishing to minimize its weight. Denoting by $u(t)$ the parameter determining the cross-section, the cost functional is

$$J(u(t)) = \int_{t=0}^{T} A(u(t))dt . \tag{1.1}$$

The coordinates chosen here are: t - the distance from the free end measured along the shape of the column; x - the lateral deflection. We denote by $u(t)$ a design parameter, $A(u(t))$ - the cross-sectional area, $I(u(t))$ - the smallest moment of inertia of the cross-sectional area about a centroidal axis. The equation of static equilibrium is

$$EI(u) \frac{d^2x}{dt^2} + P \cdot x = 0 . \tag{1.2}$$

$x(t)$ satisfies boundary conditions

$$x(0) = 0 , \quad \frac{dx(T)}{dt} = 0 . \tag{1.3}$$

A condition imposed on design is

$$\frac{P}{A(u(t))} - \sigma_{max} \le 0 , \tag{1.4}$$

where σ_{max} is the greatest allowable stress.

The problem can now be restated as an optimal control theory problem, by relabelling

$$x(t) = x_1(t) ,$$

$$\frac{dx(t)}{dt} = x_2(t) .$$

It is now reduced to a well-known formulation identical with the original control problem considered by Pontryagin et al in [1], Chapter 1.

Given:

(1.5a) $$\frac{dx_1}{dt} = x_2 = f_1(t)$$

(1.5b) $$\frac{dx_2}{dt} = \frac{Px_1(t)}{EI(u(t))} = f_2(t)$$

(1.6a) $$x_1(0) = 0$$

(1.6b) $$x_2(T) = 0$$

find a control $u(t)$, $0 \le t \le T$ such that

(1.7) $$\phi(u) = \frac{P}{A(u)} - \sigma_{max} \le 0$$

and such that

(1.8) $$J(u) = \int_0^T A(u(t))dt$$

is minimized.

Introducing dual variables λ_0, λ_1, λ_2, μ, one constructs a Hamiltonian

$$H = \{-\lambda_0 A(u) + \lambda_1 x_2 + \lambda_2 x_1 \frac{P}{EI(u)} - \mu[\frac{P}{A(u)} - \sigma_{max}]\}_{t\in(0,T)}$$

(1.9) $$+ (\lambda_1 x_1 + \lambda_2 x_2)_T - (\lambda_1 x_1 + \lambda_2 x_2)_0 \ .$$

Since $\frac{d\lambda_0}{dt} = 0$ we can choose without any loss of generality $\lambda_0 \equiv 1$. (In the engineering terminology the problem is normal.) λ_1, λ_2 obey the "canonical" system of equations

(1.10a) $$\frac{d\lambda_1}{dt} = -\frac{\partial H}{\partial x_1} = \frac{P\lambda_2}{EI(u)}$$

(1.10b) $$\frac{d\lambda_2}{dt} = -\frac{\partial H}{\partial x_2} = -\lambda_1 \ ,$$

with boundary conditions

(1.11a) $$\lambda_2(0) = 0$$

(1.11b) $$\lambda_1(T) = 0 \ .$$

Eliminating λ_1 , it only remains to solve the system

(1.12)
$$\begin{cases} \dfrac{d^2\lambda_2}{dt^2} + \dfrac{P\lambda_2}{EI(u)} = 0 \\[2ex] \lambda_2(0) = \dfrac{d\lambda_2}{dt}(T) = 0 \ , \end{cases}$$

subject to condition

(1.13)
$$\begin{cases} \text{either} \quad \dfrac{P}{A(u)} - \sigma_{max} = 0 \\[1ex] \text{or} \quad \mu = 0 \ . \end{cases}$$

Choosing $A(u) = u$, $I(u) = \alpha u^2$ (rectangular shape design!), E. J. Haug obtained profiles of optimal columns previously obtained by J. B. Keller and F. I. Niordson ([29]). Various other structural design problems are solved in [17], for example, the design of towers with guide lines. In the case of pin-jointed structures the analysis of Komkov [38] can be modified for some static designs.

In this work we wish to connect this theory with duality established in the works of Noble [25], Rall [30], Robinson [31], Arthurs [32], [33], [34], Komkov [35], [36], [37], Nashed [27] and others. A clue to this procedure is given in the preceding example of Haug. The equations (1.10a), (1.10b), (1.5a), (1.5b) with conditions (1.6a), (1.6b), (1.11a), (1.11b) correspond to a stationary point of the Lagrangian functional

(1.14)
$$L = H - \sum_{i=1}^{2} \lambda_i \dot{x}_i \qquad (\cdot = \frac{d}{dt}) \ .$$

In the examples given in [17] the static problems were reducible to a system of equations in "canonical" form

(1.15a)
$$\dot{\underset{\sim}{x}} = \frac{\partial H}{\partial \underset{\sim}{p}}$$

(1.15b)
$$-\dot{\underset{\sim}{p}} = \frac{\partial H}{\partial \underset{\sim}{x}}$$

where H is the Hamiltonian satisfying Pontryagin's maximality conditions. However, many problems of engineering systems optimization theory are difficult, or perhaps impossible to restate in the form (1.15) or in any form suitable for application of known results of control theory.

In a simple problem of optimization of a simply supported or cantilevered beam, subject to a given deflection at a fixed point, Prager made a shrewd observation concerning an application of the Betti-Castigliano deflection formula.

2. A SIMPLE STRUCTURAL PROBLEM

Optimization of a statically determinate beam. In statically determinate cases of structural analysis the bending moments are functions only of the loads applied, and of boundary conditions, and are independent of the shapes of the structure, hence of the design parameters.

Hence, if the boundary conditions are fixed and the loads applied are given a priori, the design parameters enter in a particularly simple form into the cost functional, or the constraints of optimization analysis.

This fact has been utilized very effectively by W. Prager and his associates in a series of articles, where optimization of weight, or another functional, depending on the design, was derived subject to a specified deflection at a given point. Some generalizations were given by Prager's student but they were all based on the same approach utilizing the Betti formula. See, for example, [18], [19], and [20]. The existence and uniqueness of an optimal design of a beam for the problem of Prager has been given in [21]. The subsequent discussion utilizes Duhamel's principle instead of the Betti-Castigliano formula.

In what follows we shall use standard notation. Euler-Lagrange linear assumptions of beam theory are considered correct. $M(x)$ is the bending moment caused by loads applied to the beam. $m(x,x_0)$ is the bending moment at x due to a unit load positioned at x_0. E is Young's modulus. $I(x)$ is the moment of inertia of the cross-section about the neutral axis. $A(x)$ is the cross-sectioned area. $y(x)$ denotes the deflection.

Prager has solved in [16] the specific problem of optimum weight design of a statically determinate beam subject to a given deflection at a specified point, utilizing the Betti-Castigliano formula

(2.1) $$y(x_0) = \int_0^{\ell} M(x)m(x,x_0)S(x)dx$$

(2.1a) where $$S(x) = [EI(x)]^{-1} .$$

In [21] Komkov and Coleman restate Prager's result by invoking Duhamel's principle. There exists a function $G(x,\xi)$ such that the deflection function $y(x)$ is represented in the form

(2.2) $$y(x) = \int_0^{\ell} G(x,\xi)q(\xi)d\xi$$

where $q(\xi)$ is the load applied to the beam.

$G(x,\xi)$ depends on the design of the beam and on the boundary conditions, but is independent of the load. Maxwell's reciprocal relation implies that $G(x,\xi) = G(\xi,x)$.

G is called the influence function by the engineers, the Green's function by mathematicians and physicists. For a specific design $S(x)$, Duhamel's principle can be expressed by the formula

(2.3) $$y(x) = G(S(x),x) * q(x)$$

where $*$ is the convolution product defined as usual by

(2.4) $$(f * g)(x) = \int_0^{\ell} f(x - \xi)g(\xi)d\xi \quad .$$

The formula (2.3) can be replaced by

(2.3a) $$y(x) = G(S(x),x) * \frac{d^2}{dx^2} M(x) \quad .$$

The physical meaning of $G(S(\xi),\xi)$, $\xi = x - x_0$ is the deflection at x due to a unit point load at x_0 . Hooke's law implies validity of Castigliano's formula

$$G(\xi) = \int_0^{\ell} S(x)m(\xi,x)g(\xi,x)dx \quad ,$$

$$g(\xi,x) = \frac{\partial\mu(\xi,x)}{\partial\Pi(\xi)} \quad .$$

Here $\mu(\xi,x)$ denotes the moment at x due to a point load $\Pi(\xi)$ positioned at ξ .

$$m(x,\xi) = \lim_{\Pi(x)\to 1} \mu(\Pi(x),x,\xi) \quad .$$

$$g(\xi,x) = \frac{\partial m(\Pi(\xi),x,\xi)}{\partial\Pi(\xi)} \quad .$$

Unfortunately, it is not always possible to apply rules of convolution algebra to simplify Prager's results based on the Betti-Castigliano formula. (See Appendix 1.)

The analysis of the form of "the influence function" $G(z)$ for specific boundary conditions leads to specific necessary conditions for optimization, following a simple observation. In order that a functional $J(u)$, which is Fréchet differentiable, should attain an extremal value, it is necessary that the Fréchet derivative $\frac{\partial J(u)}{\partial u}$ is equal to zero.

In case of statically determinate beam design the formulation of this necessary condition is fairly simple.

Suppose we wish to minimize the deflection at a given point x_0 , subject to a constraint $Q(u)$, where u is a scalar design variable. Since M and m are independent of design, a formula similar to Prager's can be given in the alternate form

$$\lambda_0(M(x)\ \frac{\partial S(u)}{\partial u}\ ,m(x,x_0)) + \lambda_1 \frac{\partial Q}{\partial u} = 0\ .$$

In the general statically indeterminate case, the approach of Prager and associates can not be used, and attempts at generalization had limited success. However, the optimization conditions based on Duhamel's principle are not restricted to statically determinate cases.

The difficulty lies mainly in finding "the influence function." Once G(z) is known, the optimization of deflection at a given point, or of maximum deflection, subject to constraints such as constant weight are easily formulated. (See [21], §3.) Analogous results for thin plates have been derived in [21], §4.

Generalizations applicable to cylindrical shells or to axially symmetric thin shells are routine exercises, and should be at best summarized in a short note. The important aspect of the development above is the abstract formulation which connects the optimization theory with similar theories in calculus of variations and in control theory.

One point of departure would be to exploit the necessary condition of [21] by suggesting approximate numerical schemes. The schemes which would result from some "Newton-like" procedure of converging on the zero of the Fréchet derivative have been in fact already derived independently in engineering literature by several authors (Miele, Haug, Arrora, Matsui). Instead we shall exploit the possibility of abstract formulation which is reducible either to a system of the form (1.15a), (1.15b), or to a generalization of such system, under fairly "reasonable" assumptions. All functions are assumed to be measurable. Sufficient smoothness is assumed, unless otherwise stated, whenever derivatives are computed. Other assumptions follow from physical considerations. For example, we shall assume that all Hamiltonians discussed here satisfy the boundedness condition as defined by R. T. Rockafellar ([40], Inequality 1.4), so that all functionals discussed here are well-defined. The basic setting is of Sobolĕv Space H_0^m , m depending on the specific class of optimization problems. Such assumption is generally equivalent to postulating the finiteness of the total energy of the system.

3. DUALITY IN THE SENSE OF KORN, FRIEDRICHS AND NOBLE

For the sake of convenience we shall use a Hilbert space setting. The entire exposition is easily restated in a Banach space. From purely physical point of view, it is preferable to assume that the functions considered here are points in a Banach space B , since the physical distinction between a generalized displacement and a corresponding generalized force is reflected in the mathematical setting in which elements of B model the generalized displacements, and elements of B^* (the dual of B) model the corresponding generalized forces. In a Hilbert space setting H and H^* are identified with each other and the bookkeeping of physical duality is confused.

In [25] Noble made the observation that many problems of physics have formulation $A^*Aw(x) = f(x)$, where $f(x), w(x)$ are elements of a Hilbert space H_1 . A maps H_1 into a Hilbert space H_2 , and the domain of A is dense in H_1 .

A^* is the formal adjoint of A , mapping H_2 into H_1 . Introducing a dual variable $p = Ax$, Noble observed that the corresponding system

(3.1a) $$Ax = p$$

(3.1b) $$A^*p = f(x)$$

is a generalization of the Hamilton's canonical system, with the Hamiltonian

(3.2)
$$\mathcal{H} = \tfrac{1}{2}\langle p,p\rangle_{(H_2)} + F(x) ,$$
$$\text{where } \frac{\partial F(x)}{\partial x} = f(x) \in H_1 .$$

This is, of course, the simplest possible case with boundary or initial conditions completely ignored, and with absence of constraints.

The system (3.1a), (3.1b) is equivalent to "the canonical" system

(3.3a) $$A\underset{\sim}{x} = \frac{\partial \mathcal{H}}{\partial p}$$

(3.3b) $$A^*\underset{\sim}{p} = \frac{\partial \mathcal{H}}{\partial x} .$$

If $\{\hat{\underset{\sim}{x}},\hat{\underset{\sim}{p}}\}$ satisfy (3.3a), (3.3b) then the pair $\{\hat{x},\hat{p}\}$ is the critical pair of the "Lagrangian functional"

(3.4) $$L = \tfrac{1}{2}[\langle Ax,p\rangle_{(H_2)} + \langle x,A^*p\rangle_{(H_1)}] - \mathcal{H} .$$

The inverse problem of constructing variational principles, given a set of differential equations has been considered by Vainberg [12], Nashed [7], Tonti [11]. Tonti classifying differential operators came to the conclusion that no variational principles can be constructed for certain classes of equations, for example, for the equation $f(x,\dot{x},t) = 0$. Similar statements exist concerning the dual (or complementary) variational principles.

In the case of linear operator equations the statement is very simple. The operator must be symmetric. In particular, the system $\frac{\partial \mathcal{H}}{\partial u} = Bp \quad (\in H_1)$, $\frac{\partial \mathcal{H}}{\partial p} = Au \quad (\in H_2)$, defines a dual variational formulation if and only if $B = A^*$.

This is particularly amazing when one looks for a vector $\underset{\sim}{u}$ which solves a differential equation of the form $A^*A(u)_{\underset{\sim}{x}} + D(u(\underset{\sim}{x})) = f(\underset{\sim}{x})$, where A, A^*, D are differential operators, and the order of D is lower than of A^*A , but D is not a symmetric operator. For example, the equation $-\frac{\partial^2 u}{\partial x^2} + \frac{\partial u}{\partial t} = f(x,t)$ is of that form.

4. KORN'S DUALITY FOR NON-SELF ADJOINT OPERATOR A ON A HILBERT SPACE H

Let us consider the case when $H_1 = H_2 = H$. We shall consider simultaneously the equation (1.2), i.e.,

(4.1a) $$-\frac{\partial^2 u}{\partial x^2} + \frac{\partial u}{\partial t} = 0 , \quad x \in \Omega ,$$

which can be written as

(4.2a) $$A^*Au + Du = f ,$$

and

(4.2b) $$AA^*v + D^*v = 0 ,$$

where (in this particular case) $\text{Dom}(u) = \text{Dom}(v) = \Omega$, $A = \frac{\partial}{\partial x}$, $A^* = -\frac{\partial}{\partial x}$, $D = \frac{\partial}{\partial t}$, $D^* = -\frac{\partial}{\partial t}$. u,v are considered in a $W_2^{1,1}$ setting.

We formulate a functional $L_{1,2}$ in $H \times H$: where $<,>$ is the usual $(\Omega \times (0,\infty))$ L_2-product. $L_{1,2} = \langle Au,Av\rangle + \langle Du,v\rangle - \langle f,v\rangle$ whose Fréchet derivatives are

(4.3a) $$L_{1,2_v} = A^*Au + Du - f$$

(4.3b) $$L_{1,2_u} = A^*Av + D^*v .$$

Clearly the equations (4.2a) and (4.2b) represent the existence of a critical point of $L_{1,2}(u,v)$. To reproduce the Hamiltonian formalism of Noble, we introduce the generalized momenta

$$p_1 = \frac{\partial L_{1,2}}{\partial(Av)} \quad , \quad p_2 = \frac{\partial L_{1,2}}{\partial(Av)}$$

and the Hamiltonian

$$H_{1,2} = \langle p_1, Av\rangle + \langle Au, p_2\rangle - L_{1,2} \; .$$

Then

$$\frac{\partial H_{1,2}}{\partial p_1} = Av \quad , \quad \frac{\partial H_{1,2}}{\partial p_2} = Au \; ,$$

$$\frac{\partial H_{1,2}}{\partial u} = A^* p_2 - (A^*Av + D^*v) \; , \quad \frac{\partial H_{1,2}}{\partial v} = A^* p_1 - (AA^*u + Du - f) \; .$$

The vanishing of Lagrangian derivatives $L_{1,2_u} = L_{1,2_v} = 0$ at a critical point $\{\hat{u},\hat{v},p_1,p_2\} \in (H_1 \times H_2) + (H_1 \times H_2)$ implies that u,v satisfy the canonical equations,

$$\frac{\partial H_{1,2}}{\partial p_1} = Av \tag{4.4a}$$

$$\frac{\partial H_{1,2}}{\partial p_2} = Au \tag{4.4b}$$

$$\frac{\partial H_{1,2}}{\partial v} = A^* p_2 \tag{4.4c}$$

$$\frac{\partial H_{1,2}}{\partial v} = A^* p_1 \; . \tag{4.4d}$$

Conversely starting formally with the system (4.4a), (4.4b), (4.4c), (4.4d) we could introduce the Lagrangian product $L_{1,2}$ by postulating

$$L_{1,2}(u,v,p_1,p_2) = 2\, \langle p_1,p_2\rangle - H_{1,2} \; .$$

(The usual definitions include the factor ½ in front of all quantities in $H_{1,2}$ introduced above. We have omitted it.)

The system (4.4a)-(4.4d) is therefore equivalent to the system

$$\frac{\partial L_{1,2}}{\partial u} = \phi \; , \quad \frac{\partial L_{1,2}}{\partial v} = \phi \; , \quad \frac{\partial L_{1,2}}{\partial p_1} = \phi \; , \quad \frac{\partial L_{1,2}}{\partial p_2} = \phi \tag{4.4}$$

where ϕ is the zero vector.

Hence a quadruple $\{\hat{u},\hat{v},\hat{p}_1,\hat{p}_2\}$ satisfying equations (4.4a)-(4.4d) is the critical point of the functional $L_{1,2}(u,v,p_1,p_2)$.

In the particular case of equation (1.2) the Lagrangian $L_{1,2}$ is given by

$$(4.5) \qquad L_{1,2} = \langle \frac{\partial u}{\partial x}, \frac{\partial v}{\partial x} \rangle + \langle \frac{\partial u}{\partial t}, v \rangle - \langle f, v \rangle$$

and the Hamiltonian by

$$(4.6) \qquad H_{1,2} = 2 \cdot \langle p_1, p_2 \rangle - L_{1,2} \; .$$

The above discussion indicates that the device of embedding the weak solutions of the differential equation (1.2a) or of other differential equations of the form $A^*Au + Du = f$ in a larger space permits to establish a variational principle. This line of thinking will now be utilized to directly connect engineering optimization problems with Pontryagin's maximality principle. Again we shall present our arguments only in the simplest possible case, either omitting entirely the boundary and initial conditions, or assuming the most convenient conditions for the sake of simplicity of presentation. It is not hard to offer subsequent generalizations by utilizing some results of Arthurs [32], Komkov [35], [36], or Nashed [27].

5. A REDUCTION OF A STRUCTURAL OPTIMIZATION PROBLEM TO A HAMILTONIAN FORMALISM

We consider the problem of optimization of the weight of a beam. The design parameters which can be adjusted within certain limits are $C_1(x), C_2(x), \ldots, C_m(x)$, $x \in [0,\ell]$. The admissible designs satisfy certain inequalities $\gamma_1(\underset{\sim}{C}) \le 0$, $\gamma_2(\underset{\sim}{C}) \le 0$, $\ldots$, $\gamma_x(\underset{\sim}{C}) \le 0$. $\underset{\sim}{C} \overset{\text{def}}{=} \{C_1, C_2, \ldots, C_m\}$. The beam is assumed to satisfy Lagrange's linear equation

$$(5.1) \qquad \frac{d^2}{dx^2}\left[EI(\underset{\sim}{C}(x),x)\frac{d^2w}{dx^2}\right] = f_p(x) ,$$

where f_p is a prescribed load, $EI > 0$, and $0 \le x \le \ell$. We wish to minimize a functional $J(w(\underset{\sim}{C}),\underset{\sim}{C}(x))$ which is of the form

$$(5.2) \quad J(w(\underset{\sim}{C}(x),x),\underset{\sim}{C}(x)) = \int_0^{\ell} \hat{\psi}_0(\underset{\sim}{C}(x),x))dx = \int_0^{\ell} \psi(x)dx .$$

There is usually a constraint of the form

$$(5.3) \qquad \int_0^{\ell} \phi_i(w(\underset{\sim}{C}),\underset{\sim}{C}(\underset{\sim}{x}),\underset{\sim}{x})dx \le 0 , \quad i = 1,2,\ldots,k .$$

The design $\underset{\sim}{C} = \{C_1,C_2,\ldots,C_m\}$ is a vector in a Banach space B, where the norm $\| \ \|$ of B is chosen by purely physical considerations of what is an admissible design.

The set of admissible designs is denoted by $\hat{B}$ $(\subset B)$. The physical problem of beam optimization is reduced to the finding of an admissible design $\underset{\sim}{C}(x)$, such that the corresponding deflection $w(\underset{\sim}{C})$ satisfies the equation (5.1) and the inequalities (5.3), and the functional (5.2) is minimized. In the case of a cantilevered beam $w(x) = \hat{w}(\underset{\sim}{C}(x),x)$ also satisfies the boundary conditions

$$(5.3a) \qquad w(0) = \frac{\partial w(0)}{\partial x} = 0 ,$$

$$(5.3b) \qquad \frac{\partial^2 w(\ell)}{\partial x^2} = \frac{\partial}{\partial x}\left(EI(x)\frac{\partial^2 w(\ell)}{\partial x^2}\right) = 0 .$$

The geometry is prescribed so that $I(x),A(x)$ are (uniquely defined) functions of $\underset{\sim}{C}(x)$. For the sake of simplicity and mathematical convenience we shall consider $w(\underset{\sim}{x})$ to be an element of Sobolëv space H_0^2, despite some obvious objections which may be raised on physical grounds. (Observe that f_p does not belong to any Sobolĕv space if point moments and point loads are considered to be admissible loads.) Let us denote $EI(\underset{\sim}{C}(x),(x))$ by $\hat{I}(x)$, and rewrite the equation (5.1)

$$(5.1a) \qquad \frac{d^2}{dx^2}\sqrt{\hat{I}(x)}\left(\sqrt{\hat{I}(x)}\,\frac{d^2}{dx^2}W\right) = f_p = 0 ,$$

or

$$(5.1b) \qquad A^*Aw - f_p = 0 ,$$

where A^*A is considered an operator from $H_0^2[0,\ell]$ into itself. If we write $Aw = M$ then the equation (5.1b) can be represented as a Hamiltonian system

$$(5.4a) \qquad Aw = \quad = \frac{\partial H}{\partial M}$$

$$(5.4b) \qquad A^*M = f_p = \frac{\partial H}{\partial w}$$

where

$$(5.5) \qquad H = \tfrac{1}{2}\langle M,M\rangle + \langle f_p,w\rangle .$$

The corresponding Lagrangian is

$$(5.6) \qquad L = \langle f_p,w\rangle - \tfrac{1}{2}\langle M,M\rangle = H - \langle Aw,M\rangle .$$

The minimization of

$$(5.6a) \qquad J = \int_0^\ell \psi_0(\underset{\sim}{C}(x),x)dx = \int_0^\ell \hat{\psi}(x)$$

(for a chosen design $\underset{\sim}{C}$) is handled in a manner suggested by Pontryagin's approach to an identical problem of control theory.

We introduce a function

$$(5.7) \qquad w_0(\underset{\sim}{C}(x),x) = \int_0^x \psi_0(C(\xi),\xi)d\xi .$$

Obviously

(5.7a) $$w_0(\underset{\sim}{C}(0),0) = 0$$

and

(5.7b) $$w_0(\underset{\sim}{C}(\ell),\ell) = J(\underset{\sim}{C}) .$$

Let us denote by M_0 the function

$$M_0(\underset{\sim}{C}(x),x) = \{EI(C(x))\}^{\frac{1}{2}} \frac{d^2 w_0(C(x),x)}{dx^2} ,$$

i.e.,

(5.8a) $$M_0 = Aw_0 = \sqrt{EI(\underset{\sim}{C}(x))}\ \psi_0'(x)$$

where $\psi_0'(x)$ denotes $\frac{d}{dx}(\psi_0(\underset{\sim}{C}(x),x)$. We compute

(5.8b) $$A^*M_0 = \frac{d^2}{dx^2}(\sqrt{EI(\underset{\sim}{C}(x))}\ M_0) = \frac{d^2}{dx^2}(EI(\underset{\sim}{C}(x))\psi_0'(x)) .$$

Following the analogous development of control theory we can introduce a "new" Hamiltonian

(5.9) $$H_0 = \tfrac{1}{2}\langle M,M\rangle + \langle f_p,w\rangle + \tfrac{1}{2}\langle A^*M_0,w_0\rangle = H + \tfrac{1}{2}\langle M_0,M_0\rangle .$$

Corresponding "canonical" equations are

(5.10a) $$Aw = M = \frac{\partial H_0}{\partial M} ,$$

(5.10b) $$Aw_0 = M_0 = \frac{\partial H_0}{\partial M_0} ,$$

(5.10c) $$A^*M = f_p = \frac{\partial H_0}{\partial w} ,$$

(5.10d) $$A^*M_0 = \frac{\partial H_0}{\partial M_0} .$$

The "initial" conditions are $w(0) = 0$, $w_0(0) = 0$.

Corresponding conditions at $x = \ell$ and additional "boundary" conditions for $w(\underset{\sim}{c}(x),x)$ and $M(\underset{\sim}{c}(x),x)$ depend on the support at $x = 0$ and $x = \ell$.

The minimization problem becomes the problem of minimizing $w_0(\underset{\sim}{c}(\ell),\ell)$. In the case of a beam simply supported at both ends,

$$\frac{d^2w}{dx^2}(\ell) = w(\ell) = 0 ,$$

and minimization of $\underset{\sim}{w}(\ell)$, where $\underset{\sim}{w} = w_0$ is equivalent trivially to minimization of $w_0(\ell)$. The approximation techniques developed for Hamiltonian systems by Noble, Arthurs, Nashed and Robinson are applicable to the system (5.10a) — (5.10d). Additional assumptions concerning $w_0(x)$ must be made, but generally such assumptions follow from the physical interpretation of the functional $\hat{J}(w(\underset{\sim}{C}),\underset{\sim}{C}) = J(\underset{\sim}{C})$. For example, if $J(\underset{\sim}{C})$ is the total weight, it is easy to see that $w_0(x)$ is a positive monotone increasing function of x. The following points should be made. The second variation arguments can rarely be translated from the corresponding case where the design is considered constant and we only seek a solution of a differential system w, such that $J(w)$ is minimized.

6. BEAM VIBRATION IN THE PRESENCE OF DISSIPATION

Assuming viscous dissipation of the Voigt type, the equation of motion of the beam is given by

$$\frac{\partial^2}{\partial x^2}[EI(x)\frac{\partial^2 w}{\partial x^2}] + A(x)\rho\frac{\partial^2 w}{\partial t^2}$$

$$+ \frac{\partial}{\partial t}[\frac{\partial^2}{\partial x^2}E^*I(x)\frac{\partial^2 w}{\partial x^2}] = f_p(x,t) . \tag{6.1}$$

Introducing the adjoint variable $y(x,t)$, which satisfies

$$\frac{\partial^2}{\partial x^2}[EI(x)\frac{\partial^2 y}{\partial x^2}] + A(x)\rho\frac{\partial^2 y}{\partial t^2}$$

$$-\frac{\partial}{\partial t}[\frac{\partial^2}{\partial x^2}E^*I(x)\frac{\partial^2 y}{\partial x^2}] = 0 \,, \tag{6.1a}$$

we observe that the entire discussion of Section 4 is applicable. Equations (6.1), (6.1a) are necessary conditions for the stationary behavior of the Lagrangian

$$L_{1,2}(w,y) = \int_0^T\int_0^\ell \{EI(x)\frac{\partial^2 w}{\partial x^2}\frac{\partial^2 y}{\partial x^2} - \rho A(x)\frac{\partial w}{\partial t}\frac{\partial y}{\partial t}$$

$$+ y\frac{\partial}{\partial t}[\frac{\partial^2}{\partial x^2}E^*I(x)\frac{\partial^2 w}{\partial x^2}] - yf_p(x,t)\}dxdt \,. \tag{6.2}$$

In the structural optimization problem defined by a cost functional (5.6a) we proceed as before by adjoining a new variable $w_0(x)$,

$$W_0(x) = \int_0^x \psi_0(\underset{\sim}{C}(\xi),\xi)d\xi \tag{6.3}$$

and an adjoint variable $y_0(x)$ which, in this particular case, can be chosen equal to $w_0(x)$. ($J(\underset{\sim}{C})$ is time invariant, and so is w_0 .) We can follow the arguments of Section 5 to derive a Hamiltonian system of the form (4.4a)-(4.4d).

A more general from of the equation of motion could be also considered here.

Including the Timoshenko's inertia effects in equation (5.1), we obtain a more general form:

$$\frac{\partial^2}{\partial x^2}\left[EI(x)\frac{\partial^2 w}{\partial x^2}\right] + \frac{\partial}{\partial t}\left[\frac{\partial^2}{\partial x^2}E^* I(x)\frac{\partial^2 w}{\partial x^2}\right] + A(x)\rho\frac{\partial^2 w}{\partial t^2}$$

$$- \frac{\partial^2}{\partial t^2}\left\{\left(EI(x)\frac{A(x)\rho}{k} + I_p(x)\right)\frac{\partial^2 w}{\partial x^2}\right\} + \frac{I_p(x)A(x)}{k}\frac{\partial^4 w}{\partial t^4}$$

$$(6.4) \qquad = f_p(x,t) - \frac{EI(x)}{k}\frac{\partial^2 f_p(x,t)}{\partial x^2} + \frac{I_p(x)}{k}\frac{\partial^2 f_p(x,t)}{\partial t^2} .$$

In this case the right hand side (i.e., the inhomogeneous term) itself is a function of the design, hence must be included in any optimization of design scheme. It is not obvious how to adjoin a "new" variable $w_0(x)$ in this case, attempting to restate the system of equations and the optimization condition in the "canonical form" of Hamilton, analogous to (5.10a)-(5.10d).

7. CONCLUDING REMARKS

Similarity of form of Pontryagin's formalism and the Hamiltonian formalism of optimal design theory indicates that basically the problems should have similar theoretical foundations. The difficulty which plagues the optimization of design is the dependence of the operators on the design vector. In the example (6.4), not only the differential operator but also the inhomogeneous term depend on the design vector. Some efforts to restate the problem and to arrive at a form which may be immediately interpreted by known techniques of control theory were made in a recent paper of Haug and Komkov, presently submitted to *J. Optimization Theory and Applications*. However, these results are limited and a general theory for structural optimization problems have not been developed. A very serious fault of design optimization theory is a complete absence of existence theory. One could, of course, take the abstract approach - topologize the set of admissible designs $\hat{B}$ so that $\hat{B}$ becomes a compact subset of a larger space B of abstract designs, and make certain that $J(w(\underset{\sim}{C}),\underset{\sim}{C})$, $J : \underset{\sim}{C} \to R$ is a lower semicontinuous function of $\underset{\sim}{C}$; then conclude trivially that there exists $C_0 \in B$ such that for any minimizing sequence $\underset{\sim}{C}_n \in B$, $\liminf J(\underset{\sim}{C}_n) \geq J(\underset{\sim}{C}_0)$. (Acknowledgment to Prof. Berkovitz who raised this argument in a recent lecture.) Unfortunately, the scientists interpreting this result will be somewhat unhappy with our definition of convergence and might find some serious faults with the suggested optimal design. Secondly, the assertion of lower semicontinuity of $J(w(\underset{\sim}{C}),\underset{\sim}{C})$ is hard to verify in many practical cases, since $w(\underset{\sim}{C})$ may be a solution of a differential equation $Lw = q$, where the unbounded operator L depends on $\underset{\sim}{C}$. However, this introduction should be made on grounds similar to the introduction of relaxed controls. The existence of optimal relaxed design $\underset{\sim}{C}_0$ indicates the existence of admissible "genuine" designs $\underset{\sim}{C}_n$ converging to $\underset{\sim}{C}_0$, and permits a choice of a suboptimal design, which is sufficiently close to optimal.

It is clear that in the general case Filippov's, Roxin's and Warga's results concerning the existence of optimal controls are not translatable to the theory of optimal design, and a new beginning has to be made in this direction. (For reference, see [42].) Without existence theory, necessary conditions for optimality may be an analogue of Perron's paradox. ($n = 1$ is a necessary condition for n to be the largest integer, since $n \neq 1$ implies $n^2 > n$.)

A Problem of Simultaneous Control and Design Optimization

A modest effort in the direction of finding some necessary conditions for the optimal design of a structure which is controlled by a generalized force, effecting an optimal damping of vibrations was made in [39] by Komkov and Coleman. To explain the nature of the problem we consider a specific case of a vibrating beam:

$$L(w(\underset{\sim}{s};x,t)) = \rho A(\underset{\sim}{s},x)\frac{\partial^2 w(\underset{\sim}{s};x,t)}{\partial t^2} + \frac{\partial^2}{\partial x^2}(EI(\underset{\sim}{s},x))\frac{\partial^2 w(\underset{\sim}{s};x,t)}{\partial x^2}$$

$$= u(x,t)\ , \tag{7.1}$$

where $u(x,t)$ is an admissible load (see explanation in the Appendix), obeying $\|u(x,t)\| \le 1$ for all $t \in [0,T]$. Here $\underset{\sim}{s}$ is an m-tuple depending on physical design parameters (thickness, cross-sectional area, etc.), $\underset{\sim}{s} = [s_1,s_2,\ldots,s_m]$.

Equation (7.1) can be rewritten as a pair of equations:

$$M(x,t) = -EI(\underset{\sim}{s},x)(w(\underset{\sim}{s};x,t))_{xx},$$

$$(M(x,t))_{xx} + \beta A(\underset{\sim}{s},x)(w(\underset{\sim}{s};x,t))_{tt} = u(x,t)\ ,$$

with end conditions of the type

$$\begin{matrix} w(\underset{\sim}{s};\beta,t) \equiv 0 \\ M(\beta,t) \equiv 0 \end{matrix} \quad \left\{ \begin{matrix} t[0,T] \\ \beta = 0 \ \text{ or } \ \beta = 1 \end{matrix} \right. \tag{7.2a}$$

(freely supported end)

or

$$\begin{matrix} w(\underset{\sim}{s};\beta,t) \equiv 0 \\ \dfrac{\partial w(\underset{\sim}{s};\beta,t)}{\partial x} \equiv 0 \end{matrix} \quad \left\{ \begin{matrix} t \in [0,T] \\ \beta = 0 \ \text{ or } \ \beta = 1\ , \end{matrix} \right. \tag{7.2b}$$

(cantilevered end)

and with initial conditions

$$w(\underset{\sim}{s};x,0) = w_0(\underset{\sim}{s},x)\ , \quad w_t(\underset{\sim}{s};x,0) = V_0(\underset{\sim}{s},x)\ .$$

The bending moment is explicitly independent of the design parameter $\underset{\sim}{s}$, depending only on the load $u(x,t)$ and the boundary conditions. $\underset{\sim}{s}$ obeys some design constraints

$$\phi_i(\underset{\sim}{s}) \le 0 \ , \quad i = 1,2,\ldots,k \le m \ ,$$

where no restriction is given on the magnitude of the integer m, that is, on the dimension of $\underset{\sim}{s}$, except that it is finite. We are now attempting to simultaneously optimize the control $u(x,t)$ and the parameter $\underset{\sim}{s}$ to obtain some basic sets of equations which could be used for construction of algorithms.

The set of design vectors obeying all assigned constraints $\phi_i(\underset{\sim}{s}) = 0$, or $\phi_i(\underset{\sim}{s}) \le 0$, $i = 1,2,\ldots,k$ will be denoted by S. A vector $\underset{\sim}{s} \in S$ will be called admissible. Let us consider first the problem of minimizing the total energy of a vibrating beam for a fixed design parameter $\underset{\sim}{s}$ at the time $T > 0$. According to Komkov, Pontryagin's maximality principle takes the form

$$\int_0^{\ell} \left(-u(x,t)\,\frac{\partial w_H(\underset{\sim}{s};x,t)}{\partial t}\right)dx = \max_{u\in U} \int_0^{\ell} \left(-u(x,t)\,\frac{\partial w_H(\underset{\sim}{s};x,t)}{\partial t}\right)dx \ ,$$

where $u(x,t)$ is the optimal control and U is the set of admissible controls. U is a closed bounded convex subset of a Banach space. The maximality principle is a necessary condition for the minimization of the total energy $E(T)$, subject to one of the boundary conditions (7.2a) or (7.2b) at the end points, for a fixed admissible $\underset{\sim}{s}$.

$$\hat{E}(\underset{\sim}{s};T) = \left[\int_0^{\ell} \frac{\hat{M}^2(x,t)}{2EI(\underset{\sim}{s};x)}\,dx + \tfrac{1}{2}\int_0^{\ell} \rho A(\underset{\sim}{s};x)\hat{W}_{t^2}(\underset{\sim}{s};x,t)\,dx\right]_{t=T} \ .$$

(*Note:* $\underset{\sim}{s}$ is an admissible parameter vector if s satisfies the constraint conditions $\phi_i(\underset{\sim}{s}) \le 0$, $i = 1,2,\ldots,k$.)

Minimization of $E(\underset{\sim}{s};T)$ involves perturbation of $\underset{\sim}{s}$ with respect to the optimal quantity

$$E(s;T) = \int_0^{\ell} \frac{M^2(x)T}{2E} I^{-1}(\underset{\sim}{s};x)dx + \frac{1}{2}\int_0^{\ell} \rho A(\underset{\sim}{s};x)W_{t^2}(\underset{\sim}{s};x,T)dx .$$

Such small perturbation technique is justifiable by continuous (in fact, C^{∞}) dependence of $E(\underset{\sim}{s};T)$ on the design (parameter) vector $\underset{\sim}{s}$.

We regard the maximality principle as an added constraint:

$$-\int_0^t (u(\underset{\sim}{s};x,t) \left.\frac{\partial W_H(\underset{\sim}{s};x,t)}{\partial t}\right| dx - C(t,\underset{\sim}{s}) = 0 , \forall \quad t \in [0,T] ,$$

where

$$C(t,\underset{\sim}{s}) = \max_{u \in U} \int_0^t (-u, \frac{\partial W_H}{\partial t})dx .$$

We note that a sufficient condition for the existence of at least one optimal control is only convexity of the unit ball of U (not strict convexity!) and weak compactness of the set of corresponding attainable displacements. By the lemma of Aleoglu, this is equivalent to being closed and bounded in any space which is a dual of a normed space, a condition which is clearly satisfied in our case. We also make a physically motivated assumption that the optimal parameter value of $\underset{\sim}{s}$ is attained at $\underset{\sim}{s}_0$ which is a regular point of $\langle u,W_{H_t} \rangle$ to justify the use of a Lagrangian multiplier approach. Similar assumption regarding the constraints $\phi_i(\underset{\sim}{s}) \le 0$, or $\phi_i(\underset{\sim}{s}) = 0$, $i = 1,2,\ldots,k$, can actually be checked in specific engineering cases. It turns out to be correct in the example above.

Hence, we seek a minimum of

$$E(\underset{\sim}{s},T) + \sum_i \lambda_i\phi_i(\underset{\sim}{s}) + \mu(t)[\int_0^t (u(\underset{\sim}{s};x,t) \frac{\partial W_H(\underset{\sim}{s};x,t)}{\partial t})dx + C(t,\underset{\sim}{s})] ,$$

$$\lambda_i \le 0 ,$$

in the case of inequality constraints, and $\mu \leq 0$.

Optimality of the design vector $\underset{\sim}{s}$ implies the following equality:

$$\left\langle \frac{\partial E(\underset{\sim}{s},T)}{\partial s}, \delta s \right\rangle + \sum_{i=1}^{k} \lambda_i \left\langle \frac{\partial \phi_i}{\partial s}, \delta s \right\rangle + \left\langle \frac{\underset{\sim}{s}}{\partial s} \mu(t)(C(t,\underset{\sim}{s}))\right.$$

$$\left. + \mu(t) \frac{\partial}{\partial s} (u(\underset{\sim}{s};x,t)W_{H_t}(\underset{\sim}{s};x,t)dt, \delta s \right\rangle = 0 ;$$

$\underset{\sim}{s} \in S$, where, as before, $W_H(\underset{\sim}{s};x,t)$ is the solution of the homogeneous equation, obeying for each admissible $\underset{\sim}{s}$ the unique finite conditions at $t = T$, corresponding to a minimum of $E(\underset{\sim}{s};T)$.

An Example of Application

The accuracy of conventional weapons has been studied since the time of Carnot.

The problem of optimal design seemed too hard to consider, or to even formulate mathematically.

In the analysis carried out in the past, generally "one cause at a time" was considered. For example, the effect of "balloting" would be considered, while transverse vibration and "whip effects" due to the droop are ignored. Or, the vibration of the barrel would be analysed while balloting and droop effects are ignored.

Such analysis is not even helpful in understanding the problem if the effects discussed are not independent of each other. For example, the effect of torsional vibrations on transverse motion of the barrel can not be ignored if the center of gravity of the transversely vibrating system does not coincide with the center of torsion, and the two modes of vibration are then coupled. Similarly, the Borden tube effects, and the projectile tube interaction, affect the transverse vibration and should not be studied as separate phenomena.

The main objection to such study was the theoretical difficulty in the setting up of appropriate equations of motion and of realistic boundary conditions.

The mathematical problem of defining the accuracy of a weapon can be approached from the practical point of view of "what is it that we wish to minimize?" A closer look at the problem reveals that the position, angle and velocity of the barrel at the time when the projectile leaves the barrel are the most important parameters in determining the accuracy of the weapon. For example, let $u(x,t)$ denote the displacement of the gun barrel $x \in [0,\ell]$, $t \geq 0$, where x denoted the position along the gun barrel and t is time. If $x = 0$ is the muzzle, we wish to minimize

$$\left[\alpha \left\| \frac{\partial u}{\partial x} \right\| + \beta \left\| \frac{\partial u}{\partial t} \right\| \right] , \quad \left\{ \begin{array}{l} x = 0 \\ t = T \end{array} \right.$$

where T is the predicted time of the projectile leaving the muzzle. Since T can be predicted only with some probability p , we can introduce a Gaussian distribution $\psi(t)$ centered at $t = T$ and suggest a Sobolĕv norm

$$\|u\|_a^2 = \int_{x=0}^{x=B} \int_{T-\varepsilon}^{T+\varepsilon} \psi(t)[\alpha(u_x(x,t))^2 + \beta u_t(x,t)^2]dxdt ,$$

where $0 < \beta < \ell$ is determined from data. Alternately, a weight function $\eta(x)$ can be introduced

$$\int_{x=0}^{x=\ell} \eta(x)dx = 1 , \text{ and } \eta(x) > 0 ,$$

η attaining a maximum at $x = 0$.

$$\|u\|_a^2 = \int_0^{\ell} \eta(x) \int_{T-\varepsilon}^{T+\varepsilon} \psi(t)[\alpha u_x^2 + \beta u_t^2]dtdx \ .$$

The problem of predicting and improving accuracy can be reduced to the problem of finding a constrained minimum of $\|u\|_{acc}$ is the Sobolëv space H_0^2 .

This concept of accuracy is not necessarily the last word in mathematical modelling of weapon systems and certainly the bootstrap techniques described by the author elsewhere can be applied in an effort to find α,β or even to revise the whole concept of what is an accuracy, and how to represent it by a norm in a Hilbert space or perhaps only in a normed space, or by an entirely different mathematical formulation. Instead of assuming a specific set of differential equations of motion and a time-independent set of boundary conditions, the investigation could rely on energy techniques, using some recent results developed by the author.

If one assumes the Euler-Lagrange (linear) beam theory and Voigt-type dissipation, the kinetic and potential energy are easily identified. To represent the equations of motion and the adjoint equations the following bilinear products are introduced

$$T(u,v) = \tfrac{1}{2} \int_0^{\ell(t)} [\rho A(x) \frac{\partial u}{\partial t} \frac{\partial v}{\partial t}]dx$$

$$V(u,v) = \tfrac{1}{2} \int_0^{\ell(t)} (EI(x) \frac{\partial^2 u}{\partial x^2} \frac{\partial^2 v}{\partial x^2})dx$$

$$D(u,v) = \tfrac{1}{2} \int_0^{\ell(t)} E^* I(x)\left[\frac{\partial}{\partial t}\left(\frac{\partial^2 u}{\partial x^2}\right) \cdot \frac{\partial^2 v}{\partial x^2} - \frac{\partial}{\partial t}\left(\frac{\partial^2 v}{\partial x^2}\right) \cdot \frac{\partial^2 u}{\partial x^2}\right]dx \ .$$

Additional energy terms involving effects of the projectile, for example,

$$\int_{x=a_0}^{x=a_0+\ell_0} \hat{\eta}(x) \frac{\partial u}{\partial t} \frac{\partial v}{\partial t} \, dx \ , \quad \ell_0 = \ell_0(t)$$

describing the motion of the projectile, etc. can be added to kinetic, potential or dissipation terms in formulation of the Lagrangian functional

$$L_{1,2}(u,v) = T(u,v) - (V(u,v) + D(u,v)) \ .$$

The preceding discussion of Sections 6 and 7 of this paper at least points out the way to write necessary conditions for optimality of design, and to develop corresponding numerical algorithms.

A (Fortran) computor program based on the necessary conditions has been developed by the author.

REFERENCES

1. Pontryagin, L. S., V. G. Boltyanskii, R. V. Gamkrelidze and E. F. Mishchenko. "The Mathematical Theory of Optimal Processes." Interscience Publ. Co., New York, 1962.

2. Pontryagin, L. S., V. G. Boltyanskii, R. V. Gamkrelidze and E. F. Mishchenko. The theory of optimal processes; the maximum principle, *Amer. Math. Soc. Transl. 18*(1961), pp. 341-382.

3. Butkovskii, A. G. and A. Ya. Lerner. Optimal controls with distributed parameter, *Dokl. Akad. Nauk S.S.S.R. #134* (1960), pp. 778-781.

4. Butkovskii, A. G. and A. Ya. Lerner. Optimal control systems with disturbed parameters, *Avtomat. i Telemeh. #6, 21*(1960), pp. 682-691.

5. Butkovskii, A. G., A. Ya. Lerner and L. N. Poltavskii. Optimal control of a distributed oscillatory system, English translation: *Automat. Remote Control 26*(1965), pp. 1835-1848.

6. Russell, D. L. Linear symmetric hyperbolic systems, *SIAM J. Control* *4*(1966), pp. 276-294.

7. Russell, D. L. Optimal regulation of linear symmetric hyperbolic systems with finite dimensional controls, *M.R.C. Technical Report #566*, 1965, Madison, Wisconsin.

8. Russell, D. L. A classification of boundary conditions in optimal control theory of elastic systems, "Proceedings IFAC Symposium on Control Theory," Banff, Canada, 1972.

9. Komkov, V. A note on the vibration of thin inhomogeneous plates, *ZAMM 48*(1968), pp. 11-16.

10. Komkov, V. The optimal control of a transverse vibration of a beam, *SIAM J. Control #3, 6*(1968), pp. 401-421.

11. Komkov, V. The optimal control of vibrating thin plates, *SIAM J. Control #2, 8*(1970), pp. 273-304.

12. Komkov, V. "Optimal Control Theory for the Damping Vibrations of Simple Elastic Systems." Lecture Notes in Mathematics, Springer-Verlag Publ. Co., New York, 1972.

13. Komkov, V. "Optimal Control Theory for the Damping of Vibrations of Elastic Systems." Translation and Foreword by Selezov, Mir, Moscow, 1975. (Russian translation, revised and corrected version of the original Springer-Verlag publication.)

14. Barnes, E. R. Necessary and sufficient optimality conditions for a class of distributed parameter control systems, *SIAM J. Control #1, 9*(1971), pp. 62-82.

15. Fattorini, H. O. Boundary control of systems, *SIAM J. Control #3, 6*(1968), pp. 349-385.

16. Fattorini, H. O. and N. Coleman. Optimality of design and sensitivity analysis of beam theory, *Internat. J. Control #4, 18*(1973), pp. 731-740.

17. Haug, E. J. "Engineering Design Handbook." Headquarters U. S. Army Material Command, AMCP-706-192 Publication, July, 1973.

18. Prager, W. Optimal thermoelastic design for given deflection, *Internat. J. Mech. Sci. 13*(1971), pp. 893-895.

19. Shields, R. T. and W. Prager. Optimal structural design for a given deflection, *J. Appl. Math. Phys.* *21*(1970), pp. 513-516.

20. Chern, J. M. and W. Prager. Optimal design of a rotating disc for a given radical displacement of edge, *J. Optimization Theory and Appl.* *(6)2*(1970), pp. 161-170.

21. Komkov, V. and N. P. Coleman. An analytic approach to some problems of optimal design of beams and plates, *Arch. Mech.* *(4)27*(1975), pp. 565-575.

22. Gurtin, M. E. Variational principles for initial value problems, *Quart. Appl. Math.* *22*(1964), pp. 252-256.

23. Herrera, I. and J. Bielak. A simplified version of Gurtin's variational principles, *Arch. Rational Mech. Anal.* *53*(1974), pp. 131-149.

24. Tonti, E. On the variational formulation for linear initial value problems, *Ann. Mat. Pura Appl.* *95*(1973), pp. 331-360.

25. Noble, B. "Complementary Variational Principles for Boundary Value Problems, I. Basic Principles," Report No. 437, Math. Research Center, University of Wisconsin, 1964.

26. Komkov, V. Critical point theory and variational principles of solid mechanics, monograph, to be published in 1978.

27. Nashed, M. Z. Differentiability and related properties of nonlinear operators: some aspects of the role of differentials in nonlinear functional analysis, "Nonlinear Functional Analysis and Applications," (L. B. Rall, ed.), pp. 103-309. Academic Press, New York, 1971.

28. Komkov, V. Another look at dual variational principles, *J. Math. Anal. Appl.* *1*(1978), to appear.

29. Keller, J. B. and F. I. Niordson. The shape of an optimal column, *J. Math. Mech.* *(5)16*(1966), pp. 433-446.

30. Rall, L. B. Variational methods for non-linear integral equations, "Symposium on Non-Linear Integral Equations," (P. M. Anselone, ed.). University of Wisconsin Press, Madison, Wisconsin, 1964.

31. Robinson, P. D. Complementary variational principles, "Non-Linear Functional Analysis and Applications," (L. B. Rall, ed.). Proceedings of an Advanced Seminar at Math. Research Center, University of Wisconsin, Academic Press, 1971.

32. Arthurs, A. M. "Complementary Variational Principles." Clarendon Press, Oxford University, Oxford, 1971.

33. Arthurs, A. M. A note on Komkov's class of boundary value problems and associated variational principles, *J. Math. Anal. Appl. 33*(1971), pp. 402-407.

34. Arthurs, A. M. Complementary variational principles and error bounds for biharmonic boundary value problems, *Nuovo Cimento B, II, 17*(1973), pp. 105-112.

35. Komkov, V. Application of Rall's Theorem to classical elastodynamics, *J. Math. Anal. Appl. (3)14*(1966), pp. 511-521.

36. Komkov, V. A generalization of Leighton's variational theorem, *J. Applicable Anal. (4)1*(1972), pp. 377-383.

37. Komkov, V. On variational formulation of problems in the classical continuum mechanics of solids, *Internat. J. of Engrg. Sci. 6*(1968), pp. 695-720.

38. Komkov, V. Formulation of Pontryagin's Principle in a problem of structural mechanics, *Internat. J. Control (3)17* (1973), pp. 455-463.

39. Komkov, V. and N. P. Coleman. Optimality of design and sensitivity analysis of beam theory, *Internat. J. Control (4)18*(1973), pp. 731-740.

40. Rockafellar, R. T. Optimal arcs and the minimum value function in problems of Lagrange, *Trans. Amer. Math. Soc. 180*(June, 1973), pp. 53-83.

41. Cesari, L. Existence theorems for weak and usual optimal solutions in Lagrange problems with unilateral constraints, I, *Trans. Amer. Math. Soc. 124*(1966), pp. 369-412.

42. Filippov, A. F. On certain questions in the theory of optimal control, *Vestnik Moscov. Univ. Mat. Astron. Ser. (2)*(1959), pp. 25-32.

43. Rockafellar, R. T. Certain integral functionals and duality, "Contributions to Non-Linear Functional Analysis," pp. 215-216. Academic Press, New York, 1971.

APPENDIX I - AN INTERPRETATION OF DUHAMEL'S PRINCIPLE

Here $w(x,\xi)$ denotes the deflection of a beam at point $x \in [0,\ell]$ due to a point load applied at $\xi \in [0,\ell]$, L is the operator $L = \frac{d^2}{dx^2}\left(S^{-1}(x)\frac{d^2}{dx^2}\right)$.

CLAIM 1 $L(w,\xi)$ is a generalized function of a single variable $x - \xi$.

PROOF (Trivial) Suppose $\bar{w}(x,\xi)$ is the deflection at x due to unit load applied at ξ , with given boundary conditions. Hence, $\bar{w}(x,\xi)$ is the solution of the equation $L(\bar{w}(x,\xi)) = \frac{d^2}{dx^2}\left[\frac{1}{S(x)} \cdot \frac{d^2\bar{w}(x,\xi)}{dx^2}\right] = \delta(x - \xi)$, δ denoting Dirac delta function. The above formula proves the claim.

CLAIM 2 Let $z = x - \xi$. Denote $L(\bar{w}(x,\xi)) = \phi(z)$. Then $\phi(z)*q(z) = q(z)$.

PROOF $\phi(z)*q(z) = L(\bar{w}(x,\xi))*q(z) = \delta(z)*q(z) = q(z)$.

OBSERVATION It is not always possible to write $w(x) = [M(x)S(x)]*m(x)$. That is, the formula Betti-Castigliano is not necessarily representable as a product in the convolution algebra. The counter example is given by a case of a simply supported beam. For the sake of convenience the length of the beam is assumed equal to one. Then $m(x,\xi) = \left.\begin{matrix} x(1-\xi) \\ \xi(1-x) \end{matrix}\right\} \begin{matrix} x < \xi \\ \xi > x \end{matrix}$.

Since $m(x,\xi)$ can not be rewritten in the form $M(x,\xi) = \bar{m}(z)$, $z = x - \xi$, the representation $W = (MS)*\bar{m}$ is not possible. On the other hand, for a cantilevered beam $m(x,\xi) = \bar{m}(x-\xi)$, and $w = [M(x)S(x)]*\bar{m}(x)$.

APPENDIX 2 - THE LEGENDRE TRANSFORMATION

In classical mechanics we encounter the following transformation:

$$p = \frac{\partial L(x,\dot{x},t)}{\partial x} , \quad H(x,p,t) = \sum_{i=1}^{n} p_i \dot{x}_i - L(x,\dot{x},t) ,$$

changing the "Langragian formalism" into the "Hamiltonian formalism." To this transformation corresponds the problem of defining what is meant by

$$\frac{\partial L}{\partial (Tx)} T : H \to H_1 , \text{ (where } H,H_1 \text{ are Hilbert spaces)} ,$$

and generalizing this concept to a Hilbert space setting. Perhaps the best starting point is the equation

A.1 $$Ax = f \quad (\in H_1) , \quad A : H_1 \to H_1 ,$$

for which we intend to establish (weak) solutions which correspond to a critical point of some functional L , or to two (or more) critical points of functionals in some spaces possibly other than H_1 . We assume that A is positive definite and that we can find a space H_2 such that $x \in D_T$, $Tx = p \in H_2$, $T^*p = f \in H_1$, D_T dense in H_1 , $T^*T = A$. Of course, the choice of H_1 is non-unique, as can be seen by studying even the simplest examples

$$(-\frac{d^2}{dx^2} : H^2(R) \to L_2(R)) , \quad -\frac{d^2}{dx^2} = T^*T ,$$

$$T = i\frac{d}{dx} , \quad T^* = i\frac{d}{dx} , \quad T : H^2(R) \to H^1(C) ,$$

$$T^* : H^1(C) \to L_2(R) , \text{ or } T = \frac{d}{dx} , \quad T^* = -\frac{d}{dx} ,$$

$$T : H^2(R) \to H^1(R) , \quad T^* : H^1(R) \to L_2(R) ,$$

showing that A can be factored thru different Hilbert spaces H_1 and H_1' . More involved examples can be offered of multiple factoring process with corresponding multiple variational principles, in which the choices of intermediate Hilbert spaces are not uniquely determined.

We shall now consider the effect of the vector $Tx \in H_1$ on the value of the Lagrangian functional L . Regarding p as a fixed vector in H_1 , we consider $L(x,y)$ as functional, mapping the pair $x \in H_1$, $y \in H_2$ into the real line, where $y = Tx$.

L_y denotes the gradient of L (whenever it exists) restricted to H_2 . ($x \in H_1$ is ignored.) Similarly L_x denotes the gradient of L in H_1 . It is a straightforward computation that with p regarded as fixed, and with $y = Tx$, that

(A.2) $$L_y = p ,$$

and that

(A.3) $$\langle Tx,p\rangle - L = \tfrac{1}{2}\langle p,p\rangle + (f,x) = W$$

defines a new functional $W(x,p)$ satisfying

(A.4) $$\begin{cases} W_x = f \\ W_p = p \end{cases} .$$

The notation is standard. W_x is the gradient of W restricted to the space H_1, and W_p is the gradient of W restricted to the space H_2.

We shall describe the relations (A.1), (A.3) as the Legendre transformation. (See [6] for a classical definition.) The equations (1.4) are called (Hamilton's) canonical equations.

For variational formulations corresponding to (A.4) see Arthurs' exposition [32].

We shall point out that variational principles of [32] and [28] lead to more general results if the operator A in (A.1) can be written in the form $A = TSS^*T^*$ or TT^*SS^*. In that case, four variational principles can be established. Further decomposition of the operator A leads to additional variational principles. The examples which follow show that operators which occur in differential equations of nonlinear elasticity, in linear plate, and shell theory allow such decomposition. This leads to a discovery of multiple critical points of associated functionals. The theoretical discussion leads to a repetition of arguments given by Nashed, or Robinson (see [31]). The physical interpretation of these results is not trivial, and similar arguments should be developed for the general shell theory and in other related fields of inquiry.

Some examples of application of multiple critical point theory are given in [26].

SOME MINIMUM PROBLEMS OFF THE BEATEN PATH

George M. Ewing

The University of Oklahoma
Norman, Oklahoma

1 INTRODUCTION

This is a descriptive review of certain types of problems that have interested the author. They are diverse but all were motivated by potential applications. That they all involve monoteneity provides a unifying thread. Each is concerned with the existence and characterization of a point-function y_0 in a suitable family $\mathcal{Y}$ of such functions that will furnish a least value $J(y_0)$ of a functional J in competition with all y in $\mathcal{Y}$. They are problems of the calculus of variations or from optimal control theory if one prefers but they involve features not treated or at least not fully treated in the textbooks.

ISBN 0-12-632250-3

2 OPTIMAL BURNING PROGRAMS FOR ROCKET-PROPULSION AND SIMILAR PROBLEMS

The rocket pioneer R. H. Goddard proposed in 1919, [2.4], as an unsolved problem of the calculus of variations, that of finding the program for burning the fuel so as to maximize the summit height attained by a vehicle that is propelled directly away from the earth. G. Hamel, [2.5], published a note on this problem in 1927, and H. S. Tsien with R. C. Evans published a longer article, [2.11] in 1951. These much cited works are sketchy consisting largely of conclusions from a formally derived Euler equation.

Various versions of the Goddard problem, as it became known, and of similar problems for more complex maneuvers were partially treated in quite a flood of papers during the 1950's and 1960's, the majority of these in physical or engineering journals or in company reports. With rare exceptions this work drew conclusions from one or more necessary conditions usually without the benefit of a full or precise statement of the problem considered. The present writer was introduced to the subject by this literature and is particularly indebted to various works of D. F. Lawden, G. Leitmann and Angelo Miele, of which [2.6], [2.7], [2.8] are a sample.

The simplest nontrivial formulation of a Goddard problem idealizes the vehicle as a particle that starts from rest at ground level and moves straight up along a y-axis from a flat stationary earth with air but no winds. One first thinks of the initial-value problem

$$\begin{aligned} &m\dot{v} = -c\dot{m} - D(v,y) - mg \ , \quad \dot{y} = v \ , \\ &\dot{m} \le 0 \ , \ y(0) = v(0) = 0 \ , \end{aligned} \tag{2.1}$$

in which $-c\dot{m}$ is the *thrust* with $c > 0$, $-D(v,y)$ is the *drag*, and $-mg$ is the *total weight* (fuel plus structure), all relative to an upward-directed y-axis. The further boundary conditions, $0 < m(t_b) < m(0)$ involving the burnout mass and initial mass must be adjoined to (2.1).

It was recognized early that, if the drag (retarding force of the air) $D(v,y)$ were omitted from (2.1), then the maximal summit height is realized if all the fuel is burned implusively at take-off. The motion thereafter is then the same as that of a projectile of mass $m(t_b)$ fired from a gun.

It is to be noted that the following problems are all equivalent (or reciprocal each to each) in that the burning program $m(t)$ that solves a problem of any one of the three types will solve a corresponding problem of each of the other types. Let T denote the time of flight from take-off at $t = 0$ to the time T at which the summit height is reached. The three problems are:

(P_1) *Given* $m(0)$ *and* $m(T)$, *to maximize* $y(T)$.

(P_2) *Given* $m(0)$ *and* $y(T)$, *to maximize* $m(T)$.

(P_3) *Given* $m(T)$ *and* $y(T)$, *to minimize* $m(0)$.

The problem (P_1) is most like the original Goddard problem in [2.4]. Goddard surprisingly introduces continuous staging in his own formulation. The structure diminished along with the burning fuel. If this were modified to a rocket composed entirely of fuel, we would obtain the phenomenon at burnout of nothing moving at infinite velocity, a thought that should have fascinated a medieval philosopher.

A slightly generalized version of (P_3) was rather thoroughly analyzed by W. R. Haseltine and the author, [2.2]. In the light of the no-drag case mentioned above and various partial treatments of various versions of the problems (P_i) one could anticipate the likely need of an implusive burning of part of the fuel at the start. Whether additional impulsive burnings at later times might provide an advantage had been an open question. The literature had ignored this question by assuming that the motion for $0 < t \leq t_b$ = *burnout-time* was in accord with an Euler equation that yielded a continuous velocity $v(t)$.

It therefore seemed desirable to admit general nonincreasing functions m to the competition and to investigate the problem

$$m(0-) = \textit{global minimum} \tag{2.2}$$

subject to the side-conditions (system-conditions)

$$\begin{cases} m(t)\exp\left[\dfrac{v(t)+gt}{c}\right] = m(0-)\exp\left[\dfrac{v(0-)}{c}\right] \\ \qquad\qquad - \dfrac{1}{c}\displaystyle\int_0^t D[v(\tau),y(\tau)]\exp\left[\dfrac{v(\tau)+g\tau}{c}\right]d\tau \ , \\ y(t) = \displaystyle\int_0^t v(\tau)d\tau \ , \textit{ and } m(t) \textit{ is nonincreasing}, \end{cases} \tag{2.3}$$

and to the boundary conditions

$$\begin{aligned} &v(0-) = 0 \ , \quad y(0) = 0 \ , \quad v(T) = V \geq 0 \ , \\ &y(T) = Y > 0 \ , \quad m(T) = M > 0 \ . \end{aligned} \tag{2.4}$$

In the event that v and m are suitably differentiable the differential conditons (2.1) can be recovered from (2.3). Under the formulation (2.2), (2.3), (2.4), when m has a downward jump then v has an upward jump. That the assigned terminal velocity V can be > 0 is the slight generalization mentioned above, the terminal values V and Y are fixed but, of course, T is not. In the language of control theory, m is the control while y and v are state variables. In contrast with many control formulations the control has an assigned boundary value $m(T) = M$. If we revert for a moment to the inadequate differential formulation (2.1), both the control and its derivative appear, in contrast with typical control formulations.

The existence of an optimal control m_0 is proved under mild hypotheses on the drag $D(v,y)$ using the Helly selection theorem for a sequence of uniformly bounded monotone functions. The major concern was to characterize the optimal program. To do so and to show that the optimal program is unique required some further hypotheses on $D(v,y)$. The results can be described under three cases.

If $D(y,v) \equiv 0$ or if $D(v,y)$ and the assigned terminal velocity V are both sufficiently small, then the optimal $m_0(t)$ jumps from $m(0-)$ to $m(0+) = M$ and the vehicle coasts in free flight to the assigned terminal state $[v(T),y(T)] = (V,Y)$.

If the first case does not occur, and if, for a given terminal height Y , the assigned terminal velocity V is small enough, then the optimal $m_0(t)$ has an initial jump from $m(0-)$ to $M(0+) < M$ followed by continuous burning during an interval $0 < t \le t_b < T$ during which y and $\dot{y} = v$ satisfy the Euler equation based on the integrand

$$(2.5) \qquad f(t,y,\dot{y}) = (1/c)D(\dot{y},y)\exp[(\dot{y} + gt)/c]$$

that appears in (2.3). The vehicle then coasts to the assigned terminal state during the interval $t_b \le t \le T$.

If neither of the preceding cases occurs, then the minimizing $m_0(t)$ jumps at both $t = 0$ and T and decreases continuously during $0 < t < T$ in such a way that y and $\dot{y} = v$ satisfy the mentioned Euler equation.

Sufficient conditions for the global minimizing property of m_0 in the last two cases involved the construction of a Weierstrass type field associated with the integrand (2.5) that was large enough to contain all functions y that could occur under the constraints in the formulation (2.2), (2.3), (2.4).

For some other optimization problems from missilry that admit general nonincreasing mass functions m, see [2.1], [2.3].

We know that a type problem that arises in one setting frequently also occurs in other settings. The following was proposed by J. Neuringer and D. J. Newman in the October 1967 issue of the *SIAM Review*, [2.9].

"Consider the differential equation $\{D^2 - F(x)\}y = 0$, $y(0) = 1$, $y'(0) = 0$. Choose $F(x)$ subject to the conditions $F(x) \geq 0$, $\int_0^1 F(x)dx = M$ so as to maximize $y(1)$. The problem has arisen in connection with the construction of an optimal refracting medium."

Clearly F is intended to be integrable over $[0,1]$. In the event that $v = \dot{y}$ is AC on $[0,1]$ we can set $m(x) = \int_0^x F(\xi)d\xi$ and state the proposed problem in the form

$$y(1) = \int_0^1 \dot{y}(x)dx + 1 = \textit{global maximum} \tag{2.6}$$

subject to the side-conditions

$$\dot{v} = \dot{m}y, \quad \dot{y} = v, \quad \dot{m} \geq 0 \tag{2.7}$$

and to the boundary conditions

$$v(0) = 0, \quad y(0) = 1, \quad m(0) = 0, \quad m(1) = M. \tag{2.5}$$

The similarlity to a Goddard problem (P_1) suggests recasting this problem in the form

$$(2.9) \qquad y(1) = \textit{global maximum}$$

subject to the side-conditions

$$(2.10) \qquad y(x) = \int_0^x dv + 1 , \quad \int_0^x dv = \int_0^x y(\xi)dm , \qquad m \ \textit{is nondecreasing} ,$$

and to the boundary conditions

$$(2.11) \quad v(0-) = 0 , \quad y(0) = 1 , \quad m(0-) = 0 , \quad m(1) = M .$$

One finds candidates for a maximizing m_0 and the corresponding y_0 in two cases.

Case 1 $0 \leq M \leq 1$.

$$m_0(x) = \begin{cases} 0 , & x = 0 , \\ M , & x \in (0,1] , \end{cases} \qquad v_0(x) = \begin{cases} 0 , & x = 0 , \\ M , & x \in (0,1] , \end{cases}$$

$$y_0(x) = \begin{cases} 1 , & x = 0 , \\ Mx + 1 , & x \in (0,1] ; \end{cases} \text{(hence } y_0(1) = M + 1 \text{)}.$$

Case 2 $M \geq 1$.

$$m_0(x) = \begin{cases} 0 & , \quad x = 0 \quad , \\ M^{\frac{1}{2}} + Mx & , \quad x \in (0, 1 - M^{-\frac{1}{2}}] \quad , \\ M & , \quad x \in [1 - M^{-\frac{1}{2}}, 1] \quad , \end{cases}$$

$$v_0(x) = \begin{cases} 1 & , \quad x = 0 \quad , \\ M^{\frac{1}{2}} \exp M^{\frac{1}{2}} x & , \quad x \in (0, 1 - M^{-\frac{1}{2}}] \quad , \\ b & , \quad x \in [1 - M^{-\frac{1}{2}}, 1] \quad , \end{cases}$$

and $$y_0(x) = \begin{cases} 1 & , \quad x = 0 \quad , \\ \exp M^{\frac{1}{2}} x & , \quad x \in (0, 1 - M^{-\frac{1}{2}}] \quad , \\ a + bx & , \quad x \in [1 - M^{-\frac{1}{2}}, 1] \quad , \end{cases}$$

with b in the third expression for $v_0(x)$ and with a and b in the third expression for $y_0(x)$ given by the equations

$$a = (2 - M^{\frac{1}{2}})\exp(M^{\frac{1}{2}} - 1) \ , \quad b = M^{\frac{1}{2}}\exp(M^{\frac{1}{2}} - 1)$$

and hence $y_0(1) = a + b = 2\exp(M^{\frac{1}{2}} - 1)$.

The writer did not find a way to establish the suspected maximizing property of the functions m_0 shown above.

There shortly appeared a solution, [2.12], of the preceding problem by J. E. Wilkins, Jr. in which he showed that, to each F that is integrable over $[0,1]$, there corresponds a unique y with an AC derivative that satisfies the equation $\ddot{y} = F(x)y$ a.e. on $[0,1]$. He identified $M + 1$ and $2\exp(M^{\frac{1}{2}} - 1)$ as upper bounds of $y(1)$ in the two cases and showed that they are least upper bounds by constructing sequences y_n admissible under his formulation such that $y_n(1)$ tends to the stated upper bound for each of the cases.

A further step in this story in the paper [2.10] of W. T. Reid, in which he solves a more general maximum problem by introducing solutions of a suitably generalized linear second order differential equation and obtaining the unique minimizing character of the result with the aid of some of his earlier work. The solution of Neuringer and Newman problem is then exhibited as a special case.

3 ANOTHER CLASS OF PROBLEMS INVOLVING MONOTONEITY

These problems are met in a statistical setting. They do not involve a system of integral or generalized differential equations; only an integral to be minimized subject to a monotoneity constraint. The original discrete version is the easier to describe.

Suppose given, at a point directly over the origin of an x-axis, the source of an effect-producing radiation. Randomly chosen individuals from a population are placed at discrete positions x_i , $0 \le x_1 < x_2 < \cdots < x_n$ on the axis. If an individual exhibits a specified response, this is termed a success. If he/she does not it is a failure.

Consider $a_i + b_i$ observations at the same position x_i , of which a_i are successes and b_i are failures. Let $p_i = p(x_i)$ and $q_i = 1 - p_i$ denote the a priori probabilities of success and failure. The nature of the given radiation is such that $p_1 \ge p_2 \ge \cdots \ge p_n$. The a priori probability that, for each integer i , $1 \le i \le n$, a particular a_i of the trials will result in success and b_i in failures is then the product $\Pi\, p_i^{a_i}(1 - p_i)^{b_i}$. The maximum likelihood estimate of the unknown probabilities p_i is an n-tuple $\bar{p} = \{\bar{p}_1,\ldots,\bar{p}_n\}$ that is a solution of the problem

$$\prod_1^n p_i^{a_i}(1 - p_i)^{b_i} = \textit{maximum} \tag{3.1}$$

subject to the constraint that

$$1 \geq p_1 \geq p_2 \geq \cdots \geq p_n \geq 0 \ . \tag{3.2}$$

In order to exhibit the solution of this problem define, for each pair (r,s) of integers,

$$\begin{aligned} &\alpha(r,s) = \sum_{\nu=r}^{\nu=s} a_\nu \ , \quad \beta(r,s) = \sum_{\nu=r}^{\nu=s} b_\nu \ , \\ &\text{and} \quad A(r,s) = \frac{\alpha(r,s)}{\alpha(r,s) + \beta(r,s)} \ . \end{aligned} \tag{3.3}$$

It is shown in [3.1] that the unique maximizing $\bar{p}$ has the components

$$\begin{cases} \bar{p}_i = \min\limits_{1\leq r\leq i} \ \max\limits_{i\leq s\leq n} \ A(r,s) = \max\limits_{i\leq s\leq n} \ \min\limits_{1\leq r\leq i} \ A(r,s) \\ \quad = \min\limits_{1\leq r\leq i} \ \max\limits_{r\leq s\leq n} \ A(r,s) = \max\limits_{r\leq s\leq n} \ \min\limits_{1\leq r\leq i} \ A(r,s) \ . \end{cases} \tag{3.4}$$

It is possible, as one may suspect from the form of the expressions (3.4), to construct a game of which $\bar{p}$ represents the value. However, such a game is an afterthought that made no contribution to solving the problem.

One next wonders about the nature of a solution of a continuous analogue of problem (3.1), (3.2). Consider an arbitrary probability distribution over a closed interval $[0,X]$. Proceeding heuristically let $\alpha(x_i)$ denote the observed fraction of individuals on the i-th subinterval of $[0,X]$ generated by a partition that exhibit the specified response. This leads to a problem of a similar pattern to (3.1), (3.2), namely,

$$(3.5)\quad \prod_{1}^{n} [p(x_i)]^{\alpha(x_i)(\Delta\mu)_i} [1 - p(x_i)]^{[1-\alpha(x_i)](\Delta\mu)_i} = \textit{maximum}\ ,$$

in which $(\Delta\mu)_i$ is a measure of the number of individuals on the i-th subinterval, subject to the constraints that

$$(3.6)\quad \begin{array}{l} 1 \geq p(x) \geq 0 \ \textit{and that}\ p(x) \\ \textit{be nonincreasing on}\ [0,X]\ . \end{array}$$

If we replace the product (3.5) by the negative of its logarithm and pass to the limit as the norm of the partition of $[0,X]$ tends to 0 we have the problem

$$(3.7)\quad \int_{[0,X]} -\{\alpha(x)\log p + [1 - \alpha(x)]\log(1 - p)\}d\mu = \textit{minimum}$$

on the family M of all functions $p : [0,X] \to R$ that satisfy the conditons (3.6).

The preceding discrete version is recovered if μ is the measure generated by a suitable step function. The probability $p(x)$ in (3.6) can be interpreted as an a priori probability $\Pr\{\underset{\sim}{x} \geq x\}$ in which $\underset{\sim}{x}$ is the random variable, maximum coordinate x at which an individual from the given population will exhibit the specified response to the given stimulus. Hence a minimizing function $\bar{p}(x)$ for the problem (3.7), (3.6) provides a maximum likelihood estimate of the distribution function $F(x) = 1 - p(x)$ of the mentioned random variable.

H. D. Brunk, W. T. Reid and the writer obtained the following results analogous to (3.4) for the problem (3.7), (3.6) but the only published report was an abstract, [3.3]. Let I be a generic symbol for a subinterval of $[0,X]$. Such an interval I may include either, neither, or both of its end points u and v. Define

$$(3.8)\quad A(I) = \int_I \alpha(x)d\mu\ ,\quad M(I) = A(I)/\mu(I)\ \ \textit{if}\ \ \mu(I) \neq 0\ ,$$

and

$$(3.9) \quad p^*(x) = \inf_{0\le u\le x} \sup_{x\le v\le X} M(I) , \quad p_*(x) = \sup_{x\le v\le X} \inf_{0\le u\le x} M(I) .$$

It was found that $p^*(x) \ge p_*(x)$ a.e. (μ) on $[0,X]$, and that there exists a minimizing function $\bar{p}$ for the problem (3.7), (3.6) that is unique in the sense that any minimizing function is equal a.e. (μ) to both p^* and p_* .

H. D. Brunk, W. R. Utz and G. M. Ewing extended, [3.4], many of the preceding results to admissible functions from R^n to R of two types: (1) admissible functions that are isotone in certain of the independent variables x_i and antitone in each of the others; (2) admissible functions with this and the additional property that the second-order differences in each pair (x_i, x_j) , $i,j = 1,\ldots,n$ are either ≥ 0 or ≤ 0 . Representations of the forms (3.9) for the optimizing functions were obtained for multidimensional integrals with respect to a measure μ having integrands of the form in (3.7). Certain other types of integrals to which the methods apply were identified. Each of the two mentioned types of minimum problems has potential applications some of which are described in the introductory section of [3.4]. Among the things needed in [3.4] are some higher dimensional analogues of the Helly selection theorem, which are treated in the second paper listed under [3.4].

There has not been a wide interest in integral minimum problems of the sort described above. The writer has noticed a paper of J. S. Rustagi, [3.9], based on his Stanford dissertation that makes contact with this problem area. D. J. Newman proposed in 1962, [3.5], the problem of finding a least squares estimate of a given function F from $[0,1]$ to the reals that need not be monotone subject to the constraint that the estimate be nondecreasing. This is the problem

$J(p) = \int_0^1 [p - F(x)]^2 dx =$ *minimum* , a special case under the results of [3.4].

Two papers [3.7] and [3.8] of W. T. Reid are in contact with this problem area.

H. D. Brunk singly and with various collaborators has pushed forward the theory and application to mathematical statistics of the discrete minimum problems, a program covered by the recent book, [3.2].

4 APPROXIMATING A PREASSIGNED MAGNETIC FIELD

The version of the raw problem to be considered here is the following. How should a circular cylinder of radius a and half-length L be wound with wire to carry an electric current I so that the axial magnetic field along the axis of the cylinder will be as nearly constant as possible? Such problems are encountered in the design of a system to measure magnetic properties of a long object to be placed inside the required current carrying coil in a coaxial position.

If an electric current I flows around a circular path of radius a in the plane pierced perpendicularly by the x-axis at the position ξ , then, at an arbitrary x on that axis, there is a magnetic field intensity of magnitude $h(x - \xi)$ in the direction of the axis with the sense indicated by the right hand rule. Moreover,

$$h(x - \xi) = \frac{2\pi I a^2}{[a^2 + (x - \xi)^2]^{3/2}} \quad . \tag{4.1}$$

Let n denote the *cumulative turn function* for a circular helix of radius a about the x-axis, $-L \le x \le L$; i.e., $n(x)$ is the number of turns on the interval $[-L,x]$ with $n(-L) = 0$ and $n(L) = 2N$, the total number of turns in the helix.

It is assumed that the radius of the conducting wire is much smaller than the radius a of the cylinder of half-length L that is to carry the helix. There is then essentially no loss in generality if we suppose at the outset that the pitch of the helix is everywhere small enough so that each turn can be regarded as circular. This can be realized by choosing the total number of turns $2N$ to be suitably large.

It seems reasonable to require in the formulation of the mathematical problem that n be AC on $[-L,L]$ so as to set

$$(4.2) \qquad n(x) = \int_{-L}^{x} \dot{n}(\xi)d\xi \ , \quad \dot{n}(\xi) = \textit{turn-density at } \xi \ .$$

Given such an n , the magnitude $H(x;n)$ of the axial field at the position x on the axis of the helix is given by the equation

$$(4.3) \qquad H(x;n) = \int_{-L}^{L} \dot{n}(\xi)h(x - \xi)d\xi \ .$$

One's feel for the proposed problem is aided by recalling the well-known result that, if $\dot{n}(\xi)$ is a constant and $L = \infty$, then $H(x;n) = 4\pi\dot{n}I$ is a constant. The integral (4.3) with $\dot{n}(\xi)$ a constant and L finite gives a field intensity $H(x;n)$ that is even on $[-L,L]$ and strictly decreasing on $[0,L]$. Hence intuition suggests that a function n can only lead to a nearly constant $H(x;n)$ if it is nonnegative and nondecreasing on $[-L,L]$ and if its derivative $\dot{n}$ is even on $[-L,L]$ and nondecreasing on $[0,L]$. Less widely known but also obtainable by an elementary integration is the fact that, if an ellipsoid of revolution about the x-axis with semiaxis L in the x-direction and semiaxes a_0 in the y and z-directions is wound with $\dot{n}(\xi) = \textit{constant}$, then the field $H(x;n)$ is again constant, namely,

$$H(x;n) = \frac{8L}{3a_0}\pi\dot{n}I \quad \text{if} \quad -L \leq x \leq L .$$

But the proposed problem requires winding a cylinder.

The loosely stated requirement that $H(x;n)$ be as nearly constant as possible, $-L \leq x \leq L$, does not provide a unique criterion of optimality. Among possible functionals J whose infima should be useful are:

$$J(n) = \sup_{x \in [-L,L]} |H(x;n) - H_1| \tag{4.4}$$

H_1 *a preassigned constant* ,

$$J(n) = \int_{-L}^{L} [H(x;n) - H_1]^2 dx . \tag{4.5}$$

With reasonable looking conditions on a family of functions n in which to seek a minimizing member it is easy to prove the existence of a minimizing n_0 for (4.4), (4.5) and various other measures of the deviation of $H(x;n)$ from constancy. However, the proposed problem requests specifications for winding the cylinder that could be the basis for design and fabrication.

In the remarks that follow, $J(n)$ will mean that of (4.5). The right member of (4.5) suggests the least squares problem near the end of Section 3 but when the expression (4.3) is substituted into (4.5) it is clear that the problem

$$(4.6)\quad J(n) = \int_{-L}^{L} \left\{\int_{-L}^{L} [\dot{n}(\xi)h(x-\xi) - \frac{H_1}{2L}]d\xi\right\}^2 dx = \textit{minimum}$$

is of a different structure than the example in Section 3.

There is no obvious way to obtain a necessary condition that suitably indentifies a candidate for the desired minimizing n_0. Equating the first variation of (4.6) to zero yields an integral equation as forbidding as the problem (4.6).

One tempting move is to apply the CBS inequality to the inner integral of (4.6) obtaining the inequality

$$(4.7)\quad J(n) \le J_1(n) = 2L \int_{-L}^{L}\int_{-L}^{L} [\dot{n}(\xi)h(x-\xi) - \frac{H_1}{2L}]^2 d\xi dx .$$

After reversing the order of integration we have that

$$J_1(n) = \int_{-L}^{L} \left[\varphi_2(\xi)\dot{n}^2(\xi) - \frac{H_1}{L}\varphi_1(\xi)\dot{n}(\xi) + \left(\frac{H_1}{2L}\right)^2 \right] d\xi ,$$

in which

$$(4.8)\quad \varphi_1(\xi) = \int_{-L}^{L} h(x-\xi)dx \quad \textit{and} \quad \varphi_2(\xi) = \int_{-L}^{L} h^2(x-\xi)dx .$$

Consider the auxiliary problem $J_1(n) = \textit{minimum}$ subject to the initial condition $n(-L) = 0$ with the terminal value $n(L)$ left free, to be determined in the end by the preassigned constant H_1 from (4.10) below. The Euler condition and the transversality condition at L yield the necessary condition on a minimizing n_0 for J_1 that

$$(4.9)\quad \dot{n}_0(\xi) = \frac{H_1}{2L}\frac{\varphi_1(\xi)}{\varphi_2(\xi)}$$

and hence

$$n_0(\xi) = \int_0^\xi \dot{n}_0(s)ds \ . \tag{4.10}$$

Since the integrand of $J_1(n)$ is strictly convex in $\dot{n}$, the value $J_1(n_0)$ is the global minimum of $J_1(n)$ but not a sharp estimate of the desired infimum of (4.6), which surely must be zero.

Let $P = \{-\xi_m = -L,\ldots,-\xi_1,\xi_0 = 0,\xi_1,\ldots,\xi_m = L\}$ be a partition of $[-L,L]$ and consider turn-densities $\dot{n}$ that are step-functions with the values $\dot{n}(\xi) = c_1$ on $[-\xi_1,\xi_1]$ and $\dot{n}(\xi) = c_i$ on $[-\xi_i,-\xi_{i-1}] \cup [\xi_{i-1},\xi_i]$, $i = 2,\ldots,m$. Then determine the constants c_i by equating the field $H(x;n)$ of (4.3) to a constant H_1 for each of m values of x suitably dispersed on $[0,L]$. This procedure should provide a field $H(x;n)$ approximating the given H_1 within a preassigned tolerance if the norm of P is sufficiently small.

There ought to be an approach to the posed problem of this section that characterizes a function n verifiably minimizing some good measure of the deviation from constancy of the field $H(x;n)$ or that at least yields an iterative procedure converging to such a field together with an explicit estimate of the deviation from constancy of the field $H(x;n_k)$ obtained from k steps in the process. However, a number of different methods tried from time to time have failed to produce satisfactory results of either type.

REFERENCES

For Section 2 (Optimal Burning Programs and Similar Problems)

2.1 Ewing, G. M. A fundamental problem of navigation in free space, *Quart. Appl. Math.* *18*(1961), pp. 353-362.

2.2 Ewing, G. M. and W. R. Haseltine. Optimal programs for an ascending missile, *J. SIAM Control* *2*(1964), pp. 66-88.

2.3 Ewing, G. M. Thrust direction programs for maximal range, *J. Math. Anal. Appl.* *16*(1966), pp. 347-354.

2.4 Goddard, R. H. A method of reaching extreme altitudes, Smithsonian Misc. Collections (1919), reprinted by Amer. Rocket Soc. (1946).

2.5 Hamel, G. Über eine mit dem Problem der Rakete zusammenhängende Aufgabe der Variationsrechnung, *Angew. Math. Mech.* *7*(1927), pp. 451-452.

2.6 Lawden, D. F. Maximum range of intercontinential missiles, *Aeronaut. Quart.* *8*(1957), pp. 269-278.

2.7 Leitmann, G. Solution of Goddard's problem, *Astronaut. Acta.* *2*(1956), pp. 55-62.

2.8 Miele, A. Generalized variational approach to the optimal thrust programming for the vertical flight of a rocket, Part I, *Zeit. Flugwiss.* *6*(1958), pp. 69-77; with C. R. Cavoti, Part II, *Zeit. Flugwiss.* *6*(1958), pp. 102-109.

2.9 Neuringer, J. and D. J. Neuman. Problem 67-17, an extremal problem, *SIAM Rev.* *9*(1967), p. 748.

2.10 Reid, W. T. A maximum problem involving generalized linear differential equations of the second order, *J. Diff. Equations* *8*(1970), pp. 283-293.

2.11 Tsien, H. S. and R. C. Evans. Optimum thrust programming for a sounding rocket, *J. Amer. Rocket Soc.* *21*(1951), pp. 97-107.

2.12 Wilkins, J. E., Jr. Problem 67-17, an extremal problem; solution, *SIAM Rev.* *11*(1969), pp. 86-88.

For Section 3 (Another Class of Problems Involving Monotoneity)

3.1 Ayer, Miriam, H. D. Brunk, G. M. Ewing, W. T. Reid and Edward Silverman. An empirical distribution function for sampling with incomplete information, *Ann. Math. Stat.* *26*(1955), pp. 641-647.

3.2 Barlow, R. E., D. J. Bartholomew, J. M. Bremmer and H. D. Brunk. "Statistical Inference under Order Restrictions, The Theory and Application of Isotonic Regression," John Wiley, New York et al, 1975. (See pp. 365-377 of this for an extensive bibliography.)

3.3 Brunk, H. D., G. M. Ewing and W. T. Reid. The minimum of a certain definite integral suggested by the maximum likelihood estimate of a distribution function, (Abstract) *Bull. Amer. Math. Soc.* *60*(1954), p. 535.

3.4 Brunk, H. D., G. M. Ewing and W. R. Utz. Minimizing integrals in certain classes of monotone functions, *Pac. J. Math.* *7*(1957), pp. 833-847; also the prefatory paper by the same authors, Some Helly theorems for monotone functions, *Proc. Amer. Math. Soc.* *7*(1956), pp. 776-783, which overlaps in part with 3.6 below.

3.5 Newman, D. J. Problem 61-11, on a least squares approximation, *SIAM Rev.* *4*(1962), p. 48.

3.6 Nikodym, O. M. A theorem on infinite sequences of finitely additive real-valued measures, *Sem. Mat. della Univ. di Padova* *24*(1955), pp. 255-286.

3.7 Reid, W. T. A simple optimal control problem involving approximation by monotone functions, *J. Optimization Theory Appl.* *2*(1968), pp. 365-377.

3.8 Reid, W. T. A class of optimal control problems involving higher order isotone approximations, "International Conference on Differential Equations, 1975," Academic Press, New York et al, 1975.

3.9 Rustagi, J. S. On minimizing and maximizing a certain integral with statistical applications, *Ann. Math. Stat.* *28*(1957), pp. 309-329. (See the bibliography of this for other works.)

For Section 4 (Approximating a Preassigned Magnetic Field)

There is a scattered literature in the physical journals on the magnetic fields associated with various types of coils carrying electric currents but nothing that deals with minimum problems of the type described in Section 4 insofar as the writer is aware.

EXISTENCE THEORY FOR OPTIMAL CONTROL PROBLEMS

Leonard D. Berkovitz

Purdue University
West Lafayette, Indiana

I. INTRODUCTION

In this paper we present our view of the principal features of the current state of existence theory for optimal control problems. We do not intend to give a complete literature survey nor to state the sharpest possible results. Our objective is to give a broad overview of the topic, yet present sufficient detail to be informative and to orient the reader who wishes to pursue the subject in more detail. The paper in this volume by E. J. McShane [15] further motivates some of the material presented in this paper.

Preparation of this manuscript partially supported by NSF Grant MCS7507947A1, and by Army Research Office, Durham grant ARO-DAAG29-77-M-0059.

ISBN 0-12-632250-3

II. PROBLEM STATEMENT

We shall be concerned with the following optimal control problem. Minimize the functional

$$J(\varphi,u) = g(t_0,\varphi(t_0),t_1,\varphi(t_1)) \tag{1}$$
$$+ \int_{t_0}^{t_1} f^0(t,\varphi(t),u(t))dt$$

subject to the state equation

$$\frac{dx}{dt} = f(t,x,u(t)) \ , \tag{2}$$

end condition

$$(t_0,x_0,t_1,x_1) \in \mathcal{B} \ , \tag{3}$$

and control constraints, u measurable and

$$u(t) \in \Omega(t) \ , \quad \text{a.e.} \tag{4}$$

Here $\mathcal{B}$ is a pre-assigned set in E^{2n+2} , Ω is a mapping from E^1 to subsets of E^m , g is a real valued function defined on $\mathcal{B}$, f^0 is a real valued function defined on (t,x,z)-space, $E^1 \times E^n \times E^m$, and f is a mapping from (t,x,z)-space $E^1 \times E^n \times E^m$ to E^n .

A measurable function $u = (u^1,\ldots,u^m)$ defined on an interval $[t_0,t_1]$, satisfying (4), and such that there exists an absolutely continuous function φ defined on $[t_0,t_1]$ and satisfying

$$\varphi'(t) = f(t,\varphi(t),u(t)) \quad \text{a.e.}$$

$$(t_0,\varphi(t_0),t_1,\varphi(t_1)) \in \mathcal{B}$$

$$t \to f^0(t,\varphi(t),u(t)) \in L_1[t_0,t_1]$$

is called an *admissible control*. The corresponding function φ is called an *admissible trajectory*. The pair (φ,u) is called an *admissible pair*. Let $\mathcal{A}$ denote the set of all admissible pairs. In terms of the definitions just introduced the optimal control problem can be stated as follows. Minimize $J(\varphi,u)$ over the set $\mathcal{A}$.

In many problems a set $\mathcal{R}$ of (t,x) space is given and it is required that all trajectories be such that $(t,\varphi(t)) \in \mathcal{R}$; i.e., the trajectories are required to lie in $\mathcal{R}$. In such problems the obvious modifications are made in the definitions of admissible control and admissible trajectory.

In the free end point problem in the calculus of variations the following are given: a set $\mathcal{B}$ in R^{2n+2}, a real valued function g defined on $\mathcal{B}$, and a real valued function f^0 defined on (t,x,x')-space $R^1 \times R^n \times R^n$.

Let Φ denote the class of all absolutely continuous functions φ defined on intervals $[t_0,t_1]$ such that $(t_0,\varphi(t_0),t_1,\varphi(t_1)) \in \mathcal{B}$ and such that $t \to f^0(t,\varphi(t),\varphi'(t))$ is in $L_1[t_0,t_1]$. It is required to minimize

$$(5) \quad G(\varphi) = g(t_0,\varphi(t_0),t_1,\varphi(t_1)) + \int_{t_0}^{t_1} f^0(t,\varphi(t),\varphi'(t))dt$$

over all φ in Φ.

The free end point problem in the calculus of variations can be obtained as a special case of the optimal control problem by taking $f(t,x,u(t)) = u(t)$ in (2) and $\Omega(t) = R^n$ in (4).

III. THE ABSTRACT OPTIMIZATION PROBLEM

The optimal control problem, the calculus of variations problem and many other optimization problems have the following form. A set X is given, a real valued function h defined on X is given, and it is required to find an element x_0 in X such that $h(x_0) \le h(x)$ for all x in X. If such an x_0

exists then h is said to attain its minimum on X at the point x_0.

In studying an optimization problem the first question that logically arises is whether h attains its minimum on X. It is well-known that if a topology can be put on X such that X is compact in this topology and h is lower semicontinuous with respect to this topology, then h will attain its minimum on X. The conditions can be lightened to require that X only be sequentially compact and that h be sequentially lower semicontinuous. In order to better understand some of the subsequent discussions, we review the proof of the last assertion.

Let μ denote the infimum of $h(x)$ over X. Then there exists a sequence $\{x_n\}$, called a minimizing sequence, such that $h(x_n) \to \mu$. Since X is compact, there is a subsequence, which we again label as $\{x_n\}$, and a point x_0 in X such that $x_n \to x_0$. It follows from the lower semicontinuity of h that $\liminf h(x_n) \geq h(x_0)$. Hence $\mu \geq h(x_0) \geq \mu$. Therefore μ is finite and $h(x_0) = \mu$.

We point out that all that is really required to ensure that h attains its minimum on X is that any minimizing sequence be compact and that h be sequentially lower semicontinuous at any limit point of a minimizing sequence.

IV. THE VARIATIONAL PROBLEM

The program suggested above can be carried out for the variational problem and for certain classes of control problems. We first discuss the variational problem. A useful notion of convergence on Φ is that of weak L_1 convergence of the derivative functions φ' and uniform convergence of the functions φ. To avoid technical complications, let us henceforth suppose that t_0 and t_1 are fixed. If $\mathcal{B}$ is compact and g is continuous then $\varphi_n \to \varphi_0$ uniformly on $[t_0,t_1]$ implies that $(t_0,\varphi_0(t_0),t_1,\varphi_0(t_1))$ is in $\mathcal{B}$ and that $g(t_0,\varphi_n(t_0),t_1,\varphi_n(t_1)) \to g(t_0,\varphi_0(t_0),t_1,\varphi_0(t_1))$. Hence the lower semicontinuity of the functional G depends on the lower semicontinuity of the integral functional

$$I(\varphi) = \int_{t_0}^{t_1} f^0(t,\varphi(t),\varphi'(t))dt .$$

We shall henceforth assume that $g \equiv 0$ in the discussion of the variational problem.

The introduction of weak L_1 convergence of the derivatives φ' and L_1 convergence of the functions φ (rather than uniform convergence) and the proof of the lower semicontinuity of the integrals with respect to this convergence was first carried out by Morrey [17] for the more general multiple integral problems in the calculus of variations. In the case of multiple integrals the use of L_1-convergence is more appropriate than uniform convergence. Morrey's lower semicontinuity theorem was extended by various writers in various ways. Here we shall state an extension due to the present writer [4]. Extensions of the result and a discussion of its relationship to other work are also given in [4].

THEOREM 1 Let $f^0 : (t,x,w) \to f^0(t,x,w)$ be a real valued function defined on $R^1 \times R^n \times R^m$ with the following properties. For a.e. t, f^0 is continuous in (x,w) and f^0 is measurable in t for each (x,w). For each (t,x), f is a convex function of w. There exists a function ψ in $L_1[t_0,t_1]$ such that for all t in $[t_0,t_1]$ and all (x,w)

$$f^0(t,x,w) \geq \psi(t) .$$

Let $\mathcal{D}$ denote the set of functions (y,z) with y in $L_p[t_0,t_1]$, $1 \le p \le \infty$, z in $L_q[t_0,t_1]$, $1 \le q \le \infty$, and such that

$$I(y,z) = \int_{t_0}^{t_1} f^0(t,y(t),z(t))dt < \infty .$$

Let $\{(y_k,z_k)\}$ be a sequence of elements in $\mathcal{D}$ such that (i) $y_k \to y$ strongly in $L_p[t_0,t_1]$; (ii) $z_k \to z$ weakly in $L_q[t_0,t_1]$; (iii) $\liminf I(y_k,z_k) < \infty$. Then (y,z) is in $\mathcal{D}$ and

$$\liminf I(y_k,z_k) \ge I(y,z) . \tag{6}$$

Note that " (y,z) is in $\mathcal{D}$ " is a conclusion of the theorem and not an assumption. If we assume that (y,z) is in $\mathcal{D}$ and if $\liminf I(y_k,z_k) = +\infty$ then (6) automatically holds, and we have the following corollary.

COROLLARY The functional I is lower semicontinuous on $\mathcal{D}$ with respect to strong convergence of y and weak convergence of z .

Having stated conditions guaranteeing lower semicontinuity we now state conditions guaranteeing compactness of minimizing sequences.

LEMMA 1 Let Ψ be a real valued nondecreasing nonnegative function defined on $[0,\infty]$ such that $\Psi(\xi)/\xi \to \infty$ and such that $f^0(t,x,x') \ge \Psi(|x'|)$ for all (t,x,x') . Let $\{\varphi_k\}$ be a minimizing sequence for the functional (5). Then there exists a subsequence $\{\varphi_k\}$ and an absolutely continuous function φ_0 on $[t_0,t_1]$ such that $\varphi_k \to \varphi_0$ uniformly and $\varphi_k' \to \varphi_0'$ weakly in $L_1[t_0,t_1]$.

Since the functions $\{\varphi_k\}$ belong to a minimizing sequence, there is a constant C such that

$$C \geq \int_{t_0}^{t_1} f^0(t,\varphi_k(t),\varphi_k'(t))dt \geq \int_{t_0}^{t_1} \Psi(|\varphi_k'(t)|)dt .$$

Thus, for functions $\{\varphi_k\}$ in a minimizing sequence, the conditions guaranteeing that the $\{\varphi_k'\}$ have equi-absolutely continuous integrals are fulfilled (e.g., [12], Theorem 14, p. 235). Since $\mathcal{B}$ is compact, the functions φ_k are equi-absolutely continuous and uniformly bounded. Therefore there is a subsequence, which we again label as $\{\varphi_k\}$, and an absolutely continuous function φ_0 such that $\varphi_k \to \varphi_0$ uniformly and $\varphi_k' \to \varphi_0'$ weakly in L_1.

We now state a standard existence theorem for the free end point problem in the calculus of variations. It is an immediate corollary of Theorem 1 and Lemma 1.

THEOREM 2 Let f^0 be continuous and for each (t,x) let f^0 be a convex function of x'. Let $\mathcal{B}$ be compact, let g be continuous and let the set of admissible functions φ be nonempty. Let there exist a nondecreasing, nonnegative function Ψ defined on $[0,\infty)$ such that $f^0(t,x,x') \geq \Psi(|x'|)$ for all (t,x,x') and such that $\Psi(\xi)/\xi \to \infty$ as $\xi \to \infty$. Then the functional $G(\varphi)$ attains its minimum.

We emphasize that the convexity assumption guarantees the lower semicontinuity (Theorem 1) and the "growth condition," $f^0(t,x,x') \geq \Psi(|x'|)$, where Ψ is such that $\Psi(\xi)/\xi \to \infty$, guarantees the compactness.

V. CONTROL PROBLEMS WITH LINEAR STATE EQUATIONS

Control problems in which the state equations (2) are linear and in which the integrand f^0 is a convex function of z for each (t,x) can also be treated in the context of minimizing lower semicontinuous functionals on compact sets. We now suppose that the state equations (2) have the form

$$\frac{dx}{dt} = A(t)x + B(t)u(t) + h(t) ,$$

and that the function $f^0 : (t,x,z) \to f^0(t,x,z)$ appearing in (1) is convex in z for each fixed (t,x).

We first consider the case of "unbounded controls," that is, we take $\Omega(t) = E^m$. We assume that there exists a real valued, nonnegative, nondecreasing function Ψ defined on $[0,\infty)$ such that $\Psi(\xi)/\xi \to \infty$ as $\xi \to \infty$ and such that $f^0(t,x,z) \geq \Psi(|z|)$ for all (t,x,z). We assume that the matrices A and B and the vector h are essentially bounded functions on $[t_0,t_1]$. We also assume that the end set $\mathcal{B}$ is compact. The preceding assumptions are satisfied in the linear regulator problem with quadratic cost criterion in which

$$f^0(t,x,z) = \langle x,X(t)x\rangle + \langle z,R(t)z\rangle ,$$

where for each t, $X(t)$ is positive semidefinite and $R(t)$ is positive definite.

As in the variational problem, the growth condition $f^0(t,x,z) \geq \Psi(|z|)$ and $\Psi(\xi)/\xi \to \infty$ as $\xi \to \infty$ guarantees the weak L_1-compactness of the controls in a minimizing sequence $\{u_k\}$. If u is a control then the corresponding trajectory is given by the variation of parameters formula:

$$(7) \qquad \varphi(t) = \Phi(t,t_0)\{x_0 + \int_{t_0}^{t} \Phi(s,t_0)^{-1}[B(s)u(s) + h(s)]ds\} .$$

Here $\Phi(t,t_0)$ is the matrix of fundamental solutions for the homogeneous equation

$$\frac{dx}{dt} = A(t)x$$

satisfying $\Phi(t_0,t_0) = I$. It follows from (7) that if $u_k \to u_0$ weakly in L_1 , then the corresponding trajectories φ_k converge pointwise to an absolutely continuous function φ_0 that is the trajectory corresponding to the control u_0 . Moreover, the convergence $\varphi_k \to \varphi_0$ is bounded under the hypotheses we have made. Thus, we have compactness of a minimizing sequence $\{(\varphi_k,u_k)\}$ with respect to strong L_1-convergence of the trajectories and weak L_1-convergence of the controls. Since $f^0(t,x,z) \geq \Psi(|z|) \geq 0$, and since f^0 is convex in z , Theorem 1 guarantees that the integral functional in (2) is lower semicontinuous with respect to the convergence we are discussing. We note that since $\varphi_k \to \varphi_0$ pointwise, the lower semicontinuity of $J(\varphi,u)$ will follow, provided we assume that g is lower semicontinuous.

If for each t the set $\Omega(t)$ is compact and convex and the mapping $t \to \Omega(t)$ is continuous in an appropriate sense, then it can be shown that the set of all admissible controls is weakly compact in $L_2[t_0,t_1]$. The controls in a minimizing sequence $\{u_k\}$ are therefore a fortiori weakly compact. Note that in the present case no growth conditions are required on the function f^0 . The convergence of the trajectories φ_k corresponding to a sequence of weakly convergent controls u_k is as above. The semicontinuity of the functional $J(\varphi,u)$ also follows as above.

VI. NONLINEAR STATE EQUATIONS

In the general case of nonlinear state equations (2) the procedure used to establish existence must be modified. The compactness and lower semicontinuity cannot be separated but must be combined in a single concept, which is called *lower closure*.

In the nonlinear case it is required to assume that all the admissible trajectories (or at least those in a minimizing sequence) lie in a compact set $\mathcal{R}$ of (t,x)-space. In the statement of some theorems the assumption is made directly, while in others conditions are placed on the dynamics that guarantee this.

The first existence theorems for nonlinear systems were proved by Filippov [10] and Roxin [18] for the case of bounded controls. Filippov assumed that for each t the set $\Omega(t)$ is compact and Roxin assumed that for all t, $\Omega(t) = U$, a fixed compact set. Filippov also assumed that the mapping $\Omega : t \to \Omega(t)$ satisfies a certain regularity condition. This condition, as does the assumption that Ω is the constant map, guarantees that the set

$$\mathcal{D} = \{(t,x,z) : (t,x) \in \mathcal{R}, z \in \Omega(t)\} \tag{8}$$

is compact. In terms of the problem stated here, both Filippov and Roxin required that for each (t,x) the set

$$Q(t,x) = \{(y^0,y) : y^0 = f^0(t,x,z), y = f(t,x,z), z \in \Omega(t)\} \tag{9}$$

be convex. The principal steps in the proof of existence in this case are as follows.

A "zeroth" coordinate x^0 is introduced. The equation

$$\frac{dx^0}{dt} = f^0(t,x,u(t)) \tag{10}$$

is adjoined to the state equations, and the conditions

$$x_0^0 = x^0(t_0) = 0 , \quad x_1^0 = x^0(t_1) \text{ free} \tag{11}$$

are adjoined to the set $\mathcal{B}$. Let $\hat{x} = (x^0,x)$ and let $\hat{\varphi} = (\varphi^0,\varphi)$. The problem then becomes that of minimizing the functional

$$\bar{J}(\hat{\varphi},u) = \bar{g}(t_0,\varphi(t_0),t_1,\hat{\varphi}(t_1))$$
$$\equiv g(t_0,\varphi(t_0),t_1,\varphi(t_1)) + \varphi^0(t_1) \; .$$

Note that the function

$$\bar{g} : (t_0,\hat{x}_0,t_1,\hat{x}_1) \to g(t_0,x_0,t_1,x_1) + x_1^0$$

is continuous, provided we assume g to be continuous.

The compactness of $\mathcal{B}$, $\mathcal{R}$ and $\mathcal{D}$ and the continuity of f^0 guarantee that the set of all possible values $\varphi^0(t_1)$ is bounded. Hence we need only restrict our attention to $\bar{g}$ on a compact set. Hence the infinium of $\bar{J}$ is finite. If we assume f to be continuous, it follows from the compactness of $\mathcal{D}$ that the derivatives $\hat{\varphi}_k'$ of the trajectories $\hat{\varphi}_k$ in a minimizing sequence are uniformly bounded. From this and from the compactness of $\mathcal{R}$ it follows that the trajectories are uniformly equi-absolutely continuous and uniformly bounded. Hence there exists a subsequence, which we again label as $\{\hat{\varphi}_k\}$, and an absolutely continuous function φ_0 such that $\hat{\varphi}_k \to \hat{\varphi}_0$ uniformly. It is not known at this stage of the argument that $\hat{\varphi}_0$ is a trajectory of the system (2) with (10) adjoined. If we could show that $\hat{\varphi}_0$ is a trajectory, then we would have

$$\mu = \lim_{k\to\infty} \bar{J}(\hat{\varphi}_k,u_k) = \bar{J}(\hat{\varphi}_0,u_0) \; ,$$

where $\mu = \inf \bar{J}(\hat{\varphi},u)$ and u_0 is the control corresponding to $\hat{\varphi}_0$. This, of course, would show that $\bar{J}$ attains its minimum at $(\hat{\varphi}_0,u_0)$.

To show that $\hat{\varphi}_0$ is a trajectory, one first shows that for a.e. t in $[t_0,t_1]$ there is a vector $z(t)$ in $\Omega(t)$ such that

$$\hat{\varphi}_0'(t) = \hat{f}(t,\varphi_0(t),z(t)) \; , \tag{12}$$

where $\hat{f} = (f^0,f)$. To establish this, essential use is made of the condition that the sets $Q(t,x)$ are convex. In (12) the function $t \to z(t)$ is defined pointwise. To show that $\hat{\varphi}_0$ is a trajectory, it must be shown that there is a measurable function u_0 such that $u_0(t) \in \Omega(t)$ and

$$\hat{\varphi}_0'(t) = \hat{f}(t,\varphi_0(t),u_0(t)) \quad \text{a.e.}$$

The existence of such a function u_0 is guaranteed by Filippov's Lemma, first proved in [10]. This theorem gives conditions under which the following holds. If y is a measurable function defined on an interval $[t_0,t_1]$ and if for almost all t, the equation

$$y(t) = R(t,z)$$

has a solution $z(t)$, where $z(t) \in \Omega(t)$, then there is a measurable function u with the property that $u(t) \in \Omega(t)$ a.e. and

$$y(t) = R(t,u(t)) .$$

Note that the lemma does not assert that the mapping $z \to z(t)$ is measurable. For the existence theorem there is therefore no assertion that $u_k \to u_0$ in any sense. We therefore do not have compactness in the usual sense. What is established is that under appropriate hypotheses the following holds. If $\{(\hat{\varphi}_k,u_k)\}$ is a minimizing sequence then there exists a subsequence $\{(\hat{\varphi}_k,u_k)\}$ and an admissible pair $\{(\hat{\varphi}_0,u_0)\}$ such that $\hat{\varphi}_k \to \hat{\varphi}_0$ uniformly and

$$\lim_{k\to\infty} \bar{J}(\hat{\varphi}_k,u_k) = \bar{J}(\hat{\varphi}_0,u_0) .$$

L. Cesari improved the pioneering results of Filippov and Roxin in a significant way in [7], and [8]. First, Cesari allowed the constraint mapping Ω to depend on x. Thus the mapping Ω becomes $\Omega : (t,x) \to \Omega(t,x)$, and the constraint condition (4) becomes

$$u(t) \in \Omega(t,\varphi(t)) .$$

More significantly, he replaced the requirement that the sets $Q(t,x)$ be convex by the requirement that for each (t,x) the sets

$$(13) \quad Q^+(t,x) = \{(y^0,y) : y^0 \geq f^0(t,x,z), y = f(t,x,z), z \in \Omega(t,x)\}$$

be convex. Finally, Cesari allowed the sets $\Omega(t,x)$ to be closed instead of compact, thus permitting unbounded controls.

For the case of unbounded controls the regularity assumptions on the mapping Ω, the assumption that all trajectories lie in a compact set and the continuity assumptions on f^0 and f do not guarantee that the infimum of $J(\varphi,u)$ is finite. The regularity assumptions on Ω only ensure that the set $\mathcal{D}$ defined in (8) is closed. To ensure that $\inf J(\varphi,u)$ is finite we assume that for all (t,x,z) in $\mathcal{D}$, $f^0(t,x,z)$ is bounded below in a suitable way. The simplest of these assumptions is $f^0(t,x,z) \geq \psi(t)$, where $\psi \in L_1[t_0,t_1]$.

In the unbounded control case the equi-absolute continuity of the trajectories in a minimizing sequence does not follow from the other assumptions as it does in the bounded case. Various assumptions are introduced to guarantee this. One such condition is that for each $i = 1,\ldots,n$ and for every $\varepsilon > 0$ there exists a nonnegative function F^i_ε in L_1 such that for all $\{(\varphi_k,u_k)\}$ in a minimizing sequence

$$(14) \quad |f^i(t,\varphi_k(t),u_k(t))| \leq F^i_\varepsilon(t) + \varepsilon f^0(t,\varphi_k(t),u_k(t)) \quad \text{a.e.}$$

The condition (14) guarantees that the sequence of derivatives $\{\varphi'_k\}$ of elements in a minimizing sequence is weakly compact in L_1. The uniform boundedness of the sequence $\{\varphi_k\}$, however, is automatic, since we are assuming that all trajectories lie in a fixed compact set R. Thus we conclude that there is an absolutely continuous function φ_0 and a subsequence $\{\varphi_k\}$ such that $\varphi_k \to \varphi_0$ uniformly and $\varphi_k' \to \varphi_0'$ weakly in L_1.

Since

$$\mu \equiv \inf J(\varphi,u) = \lim_{k\to\infty} J(\varphi_k,\varphi_k) ,$$

existence will be established if we can show that

$$\text{(i)} \quad \varphi_0 \text{ is a trajectory}$$

$$\text{(ii)} \quad \liminf_{k\to\infty} J(\varphi_k,u_k) \ge J(\varphi_0,u_0) , \tag{15}$$

where u_0 is a control corresponding to φ_0. Since g is continuous, $\mathcal{B}$ is compact, and $\varphi_k \to \varphi_0$ uniformly, it follows that in order to establish (15)(ii) we must show that

$$\liminf_{k\to\infty} I(\varphi_k,u_k) \ge I(\varphi_0,u_0) , \tag{16}$$

where

$$I(\varphi,u) = \int_{t_0}^{t_1} f^0(t,\varphi(t),u(t))dt .$$

In [8] and [9] Cesari gives theorems that guarantee the following. If $\{\varphi_k\}$ is a sequence of absolutely continuous functions converging uniformly to an absolutely continuous function φ_0, then φ_0 is a trajectory and (16) holds, where u_0 is a control corresponding to φ_0. We point out that (16) need not hold for all controls corresponding to φ_0. The theorem asserts that there is at least one with this property. Theorems of this type are called *lower closure* theorems.

The lower closure theorems of Cesari require that the sets $Q^+(t,x)$ be closed and convex and that for every (t,x)

$$Q^+(t,x) = \bigcap_{\delta>0} \operatorname{cl} \operatorname{co} Q^+(N_\delta(t,x)) , \tag{17}$$

where $Q^+(N_\delta(t,x))$ denotes the union of the sets $Q^+(t',x')$ over all (t',x') whose distance from (t,x) does not exceed δ. Here "cl" denotes closure and "co" denotes convex hull. Condition (17) was introduced by Cesari in [8]. One also needs appropriate bounds from below on f^0 ; those used to guarantee that μ is finite suffice.

In the applications to existence problems one also has that $\varphi_k' \to \varphi_0'$ weakly in L_1 . Under this additional hypothesis, the proof of the preceding lower closure theorem can be appreciably simplified by the use of Mazur's theorem and the Condition (17) can be replaced by the weaker condition

$$(18) \qquad Q^+(t,x) = \bigcap_{\delta>0} \text{cl co}\{ \bigcup_{|x'-x|<\delta} Q^+(t,x')\} .$$

See Berkovitz [1], Bidault [6].

If we assume that $\varphi_k' \to \varphi_0'$ weakly in L_1 and that the mapping Ω is independent of x , i.e., $\Omega : t \to \Omega(t)$, then (17) or (18) hold. We still require, however, that the sets $Q^+(t,x)$ be closed and convex. We can replace (18) by a generalized Lipschitz-Hölder condition on the function $\hat{f} = (f^0,f)$:

$$(19) \quad |\hat{f}(t,x,z) - \hat{f}(t,x,z')| \leq \mu(|x-x'|)\{H(t,z) + K|x-x'|^{p-1}\} .$$

Here $p > 1$ and μ is a nonnegative function such that $\mu(\delta) \to 0$ as $\delta \to 0$ and $\mu(\delta) \leq \delta$ for δ sufficiently large. It is also assumed that for (φ_k,u_k) in a minimizing sequence there exists a $1 \leq q \leq \infty$ and a constant A such that

$$\int_{t_0}^{t_1} |H(t,u_k(t))|^q dt \leq A .$$

See Berkovitz [2], [3]. In place of (18) we can also assume that the functions $\{u_k\}$ in a minimizing sequence all lie in a fixed ball in some L_p-space. See Berkovitz [5].

We now sketch a proof of the lower closure theorem under the assumption that (18) holds and that $\varphi_k' \to \varphi_0'$ weakly. Let $\gamma = \liminf I(\varphi_k,u_k)$. Select a subsequence such that $I(\varphi_k,u_k) \to \gamma$. Since $\varphi_k' \to \varphi_0'$ weakly, it follows from Mazur's theorem that there is a subsequence $\{\psi_j\}$ of convex combinations of the φ_k that converges strongly to φ_0'. Hence, for appropriate subsequences $\psi_j \to \varphi_0'$ almost everywhere. Let $h_k(t) = f^0(t,\varphi_k(t),u_k(t))$. Now use the same convex combinations of the function h_k as was used to form the ψ_j's to form a sequence λ_j. Let $\lambda(t) = \liminf \lambda_j(t)$. It follows from Fatou's Lemma that

$$\int_{t_0}^{t_1} \lambda(t) \leq \gamma . \tag{20}$$

For each t, except possibly those in a set of measure zero, the following holds. There exists a subsequence (depending on t) such that $\lim \lambda_j(t) = \lambda(t)$, where $\lambda(t)$ is finite. From the convergence of $\varphi_j(t)$ to $\varphi_0(t)$ it follows that there exists a sequence $\delta_j \to 0$ such that the vector $(\lambda_j(t),\psi_j(t))$ belongs to the closed convex hull of the union of the sets $Q^+(t,x')$ where x' ranges over the ball $|x' - \varphi_0(t)| \leq \delta_j$. As $j \to \infty$, we have

$$(\lambda_j(t),\psi_j(t)) \to (\lambda(t),\varphi_0'(t)) .$$

Therefore, by (18), $(\lambda(t),\varphi_0'(t)) \in Q^+(t,\varphi_0(t))$. In other words, there is a $z(t)$ in $\Omega(t,\varphi(t))$ such that

$$\lambda(t) \geq f^0(t,\varphi_0(t),z(t))$$

$$\varphi_0'(t) = f(t,\varphi_0(t),z(t)) .$$

The McShane-Warfield extension of Filippov's Lemma [16] is now used to show that there is a measurable function u_0 such that $u_0(t) \in (t,\varphi_0(t))$ and

$$\lambda(t) \geq f^0(t,\varphi_0(t),u_0(t))$$

$$\varphi_0'(t) = f(t,\varphi_0(t),u_0(t)) . \tag{21}$$

The second equation in (21) asserts that φ_0 is a trajectory of the system. Integration of the inequality in (21) and the use of (20) gives

$$\liminf_{k} I(\varphi_k, u_k) \geq \int_{t_0}^{t_1} f^0(t, \varphi_0(t), u_0(t))dt ,$$

and lower closure is established.

If we assume that Ω is independent of x and that (19) holds or that all of the controls $\{u_k\}$ in a minimizing sequence lie in some ball in an L_p-space, then we can construct a sequence $\{(\sigma_j(t), w_j(t))\}$ such that for almost all t, $(\sigma_j(t), w_j(t)) \in Q^+(t, \varphi_0(t))$ and

$$\sigma_j(t) - \lambda_j(t) \to 0 , \quad w_j(t) - \psi_j(t) \to 0 .$$

For almost every t there exists a subsequence (which may depend on t) such that $(\lambda_j(t), \psi_j(t)) \to (\lambda(t), \varphi_0'(t))$. Hence since $Q^+(t, \varphi_0(t))$ is closed and since,

$$(\lambda(t), \varphi_0'(t)) = \lim(\lambda_j(t), \psi_j(t)) = \lim(\sigma_j(t), w_j(t))$$

we conclude that $(\lambda(t), \varphi_0'(t)) \in Q^+(t, \varphi_0(t))$. The proof now proceeds as in the preceding paragraph.

VII. RELAXED CONTROLS

The existence theorems that we presented required certain regularity in the behavior of the data of the problem and required that certain sets $Q^+(t,x)$ be convex. In the case of noncompact constraints reasonable growth conditions were formulated to ensure that the trajectories in a minimizing sequence were equi-absolutely continuous. All of the conditions placed on the problem can be justified in terms of applications to actual control systems except, perhaps, the requirement that the sets $Q^+(t,x)$ be convex. This requirement restricts the applications considerably.

The control problem with no assumptions on the convexity of the sets $Q^+(t,x)$ can be embedded in a problem, called the *relaxed problem*, in which the required convexity conditions do hold. Controls and trajectories in the relaxed problem are called relaxed controls and relaxed trajectories. In the relaxed problem the required convexity conditions do hold, and the existence of a minimizing relaxed control and trajectory are guaranteed. It is then hoped that by further study of the problem, such as by applying the necessary conditions, one can conclude that the relaxed minimizing control is indeed an ordinary control. We caution the reader that this assertion need not be true.

Relaxed trajectories were first introduced by L. C. Young [21] for problems in the calculus of variations. He called them generalized curves, rather than relaxed trajectories or curves. E. J. McShane [13] studied generalized curves for Bolza problems in the calculus of variations. The relaxed controls and trajectories were rediscovered in the control problem setting independently by Gamkrelidze [11] and Warga [19]. Other treatments were given by McShane [14] and L. C. Young [22].

We shall give two equivalent formulations of the relaxed problem. In both of them we shall assume that for each t the set $\Omega(t)$ is compact and that for any finite interval $[t_0,t_1]$, the union of the sets $\Omega(t)$ as t ranges over $[t_0,t_1]$ is compact. In the first formulation, a relaxed control v on an interval $[t_0,t_1]$ is a measurable function

$$v = (\tilde{u},p) = (u_1,\ldots,u_{n+2},p^1,\ldots,p^{n+2})$$

where each u_i is a control for the original problem on $[t_0,t_1]$ and each p^i is a nonnegative measurable function such that $p^i(t) \geq 0$ and $\Sigma p^i(t) = 1$ for almost all t in $[t_0,t_1]$. A relaxed trajectory is an absolutely continuous function ψ such that

$$\psi'(t) = \sum_{i=1}^{n+2} p^i(t) f(t,\psi(t),u_i(t)) .$$

The relaxed trajectory is also required to satisfy the end condition $(t_0,\psi(t_0),t_1,\psi(t_1)) \in \mathcal{B}$. The problem is to find a relaxed control v and corresponding relaxed trajectory ψ that minimize

$$\tilde{J}(\psi,v) = g(t_0,\psi(t_0),t_1,\psi(t_1)) + \int_{t_0}^{t_1} \left[\sum_{i=1}^{n+2} p^i(t) f^0(t,\psi(t),u_i(t))\right] dt .$$

The relaxed problem can be stated in the format of the problem formulated in Section Two of this survey as follows.

For any set S and any integer q we denote the q-fold cartesian product of S with itself by $[S]^q$. Let $\tilde{z}$ denote a vector in $[E^m]^{n+2}$. Thus $\tilde{z} = (z_1,\ldots,z_{n+2})$. Let

$$\Gamma = \{\pi : \pi = (\pi^1,\ldots,\pi^{n+2}), \pi^i \geq 0, \Sigma\, \pi^i = 1\}$$

$$\tilde{\Omega}(t) = [\Omega(t)]^{n+2}$$

$$\tilde{f}^0(t,x,\tilde{z},\pi) = \sum_{i=1}^{n+2} \pi^i f^0(t,x,z_i)$$

$$\tilde{f}(t,x,\tilde{z},\pi) = \sum_{i=1}^{n+2} \pi^i f(t,x,z_i) .$$

The relaxed optimal control problem is the following.

Minimize the functional

$$\tilde{J}(\psi,\tilde{u},p) = g(t_0,\psi(t_0),t_1,\psi(t_1)) + \int_{t_0}^{t_1} \tilde{f}_0(t,\psi(t),\tilde{u}(t),\pi(t))dt \tag{22}$$

subject to the state equations

$$\frac{dx}{dt} = \tilde{f}(t,x,\tilde{u}(t),\pi(t)) \tag{23}$$

end conditions

$$(t_0,x_0,t_1,x_1) \in \mathcal{B} \tag{24}$$

and control constraints

$$\tilde{u}(t) = (u_1(t),\ldots,u_{n+2}(t)) \in \tilde{\Omega}(t)$$

$$p(t) \in \Gamma \ . \tag{25}$$

It is easy to verify that in the relaxed problem the sets

$$\tilde{Q}(t,x) = \{(y^0,y) : y^0 = f^0(t,x,\tilde{z},\pi),y = f(t,x,\tilde{z},\pi),\tilde{z} \in \tilde{\Omega}(t),\pi \in \Gamma\}$$

which correspond to the sets $Q(t,x)$ for the original problem are convex. Hence, a fortiori, so are the sets $\tilde{Q}^+(t,x)$, in whose definition the condition $y^0 \geq f^0(t,x,\tilde{z},\pi)$ replaces $y^0 = f^0(t,x,\tilde{z},\pi)$. It is also easy to verify that if the original problem satisfies the various other hypotheses of the existence theorems, so does the relaxed problem. Thus we can guarantee the existence of a solution to the relaxed problem.

We now present a second formulation of the relaxed problem, which under the hypotheses we have made can be shown to be equivalent to the first formulation. This formulation permits us to obtain existence by showing that we have a lower semicontinuous functional on a compact set. We shall follow Warga [20] in our formulation.

Let I be a compact interval containing the projection of $\mathcal{R}$ on the t-axis, where $\mathcal{R}$ is the compact set in (t,x) space to which we are restricting the trajectories. By a relaxed control v we mean a mapping

$$v : t \to v(t) \equiv \mu(t, \ )$$

to the probability measures on $\Omega(t)$ such that for every continuous function p the function P defined by

$$P(t) = \int_{\Omega(t)} p(z)d\mu(t,z)$$

is Lebesgue measurable in I. Any ordinary control u can be identified with the relaxed control v_u that assigns to t the atomic measure concentrated at $u(t)$.

A relaxed admissible pair (ψ,v) is an absolutely continuous function ψ defined on I such that $(t_0,\psi(t_0),t_1,\psi(t_1)) \in \mathcal{B}$ and a relaxed control v such that for a.e. t in $[t_0,t_1]$

$$\psi'(t) = \int_{\Omega(t)} f(t,\psi(t),z)d\mu(t,z) .$$

The problem is to find a relaxed admissible pair (ψ,v) that minimizes the functional

$$\tilde{J}(\psi,v) = g(t_0,\psi(t_0),t_1,\psi(t_1)) + \int_{t_0}^{t_1} \left[\int_{\Omega(t)} f^0(t,\psi(t),z)d\mu(t,z) \right] dt .$$

The usefulness of relaxed controls as defined here results from the fact that they are a weak star compact subset in a space of linear functionals. We now elaborate on this.

Let $\mathcal{U}$ denote the union of the sets $\Omega(t)$ as t ranges over I. By assumption $\mathcal{U}$ is compact. Let $\mathcal{L}$ denote the set of functions h defined on $I \times \mathcal{U}$ such that $h(\ ,z)$ is measurable on I for each z in $\mathcal{U}$ and $h(t, \ )$ is continuous on $\mathcal{U}$ for each t in I and such that the function M defined by the formula

$$M(t) = \max\{h(t,z) : z \in \mathcal{U}\}$$

is in $L_1[I]$. The set $\mathcal{L}$ is a Banach space with norm

$$\|h\| = \int_I M(t)dt .$$

Let $\tilde{v}$ be a mapping from I to the set of finite Radon measures on $\mathcal{U}$. Thus, $\tilde{v} : t \to \nu(t, \ )$. Let $\mathcal{N}$ denote the set of such mappings with the following two properties:

(i) For every continuous function p on $\mathcal{U}$ the function $t \to \int_{\mathcal{U}} p(t)d\nu(t,z)$ is Lebesgue measurable on I.

(ii) If $|\nu(t)|$ denotes the total variation measure of $\nu(t)$ then $\text{ess sup}\{|\nu(t)|(\mathcal{U}) : t \in I\}$ is finite. Then $\mathcal{N}$ can be identified with the dual space L^* of L. Each ν in $\mathcal{N}$ is identified with the functional ν^* by the formula

$$\nu^*(h) = \int_I \left[\int_{\mathcal{U}} h(t,z)d\nu(t,z) \right] dt , \quad h \in L .$$

The set of relaxed controls is clearly a subset of L^*. The important fact for existence questions is that the set of relaxed controls is a convex subset of L^* and is compact and sequentially compact in the weak star topology of L^* (Warga [20], Theorem IV.2.1, p. 272). Also, the functional $\tilde{J}(\psi,v)$ is continuous with respect to the weak star topology. Thus, we are in the general framework of minimizing a lower semicontinuous functional on a compact set, and existence is assured.

REFERENCES

1. Berkovitz, L. D. Existence theorems in problems of optimal control, *Studia Math.* 47(1972), pp. 275-285.

2. Berkovitz, L. D. Existence theorems in problems of optimal control without property (Q) , *in* "Techniques of Optimization," (A. V. Balakrishnan, ed.), pp. 197-209. Academic Press, New York and London, 1972.

3. Berkovitz, L. D. Existence and lower closure theorems for abstract control problems, *SIAM J. Control 12*(1974), pp. 27-42.

4. Berkovitz, L. D. Lower semicontinuity of integral functionals, *Trans. Amer. Math. Soc. 192*(1974), pp. 51-57.

5. Berkovitz, L. D. A lower closure theorem for abstract control problems with L_p-bounded controls, *J. Optimization Theory Appl. 14*(1974), pp. 521-528.

6. Bidault, M. F. Quelques résultats d'existence pour des problèmes de contrôle optimal, *C. R. Acad. Sci. Paris Sér. A-B 274*(1972), pp. 62-65.

7. Cesari, L. Existence theorems for optimal solutions in Pontryagin and Lagrange problems, *SIAM J. Control 3*(1966), pp. 475-498.

8. Cesari, L. Existence theorems for weak and usual optimal solutions in Lagrange problems with unilateral constraints I, *Trans. Amer. Math. Soc. 124*(1966), pp. 369-412.

9. Cesari, L. Closure lower closure, and semicontinuity theorems in optimal control, *SIAM J. Control 9*(1971), pp. 287-315.

10. Filippov, A. F. On certain questions in the theory of optimal control, *SIAM J. Control 1*(1962), pp. 76-89: Orig. Russ. article in *Vestnik Moskov Univ. Ser. Mat. Mech. Astr. 2*(1959), pp. 25-32.

11. Gamkrelidze, R. V. On sliding optimal regimes, *Dokl. Akad. Nauk SSSR 143*(1962), pp. 1243-1245: Translated *Soviet Math. Dokl. 3*(1962), pp. 390-395.

12. Graves, L. M. "Theory of Functions of Real Variables." 2nd. Ed. McGraw Hill, New York, Toronto, London, 1956.

13. McShane, E. J. Existence theorems for Bolza problems in the calculus of variations, *Duke Math. J. 7*(1940), pp. 28-61.

14. McShane, E. J. Relaxed controls and variational problems, *SIAM J. Control* 5(1967), pp. 435-485.

15. McShane, E. J. The calculus of variations from the beginning through optimal control theory, present collection,

16. McShane, E. J. and R. B. Warfield, Jr. On Filippov's implicit function lemma, *Proc. Amer. Math. Soc.* *18*(1967), pp. 41-47.

17. Morrey, C. B., Jr. Multiple integral problems in the calculus of variations and related topics, *Univ. of California Publ. in Math. new ser.* *1*(1943), pp. 1-130.

18. Roxin, E. The existence of optimal controls, *Mich. Math. J.* *9*(1962), pp. 109-119.

19. Warga, J. Relaxed variational problems, *J. Math. Anal. Appl.* *4*(1962), pp. 111-128.

20. Warga, J. "Optimal Control of Differential and Functional Equations." Academic Press, New York, 1972.

21. Young, L. C. Generalized curves and the existence of an attained absolute minimum in the calculus of variations, *Compt. Rend. Soc. Sci. et Lettres. Varsovie, Cl III* *30*(1937), pp. 212-234.

22. Young, L. C. "Lectures on the Calculus of Variations and Optimal Control Theory." W. B. Saunders Co., Philadelphia, London, Toronto, 1969.

NECESSARY CONDITIONS FOR MINIMUM OF ORDERS ZERO AND TWO

Jack Warga

Northeastern University
Boston, Massachusetts

1. INTRODUCTION

We shall discuss certain recent results concerning two new types of necessary conditions for a restricted minimum in general optimization problems and in optimal control problems defined by ordinary differential equations. These two types of necessary conditions can be thought of as bracketing the Pontryagin maximum principle between themselves because they apply to problems defined by functions that are Lipschitz continuous respectively, twice continuously differentiable in their dependence on the state variables while Pontryagin's principle requires continuous differentiability. We refer to the corresponding conditions as being, respectively, of orders zero and two and to Pontryagin's principle as a condition of order 1 . So as to keep the

Partially supported by Grant MCS 76-06756 of the National Science Foundation.

ISBN 0-12-632250-3

exposition as simple as possible, we shall restrict ourselves to conditions involving ordinary minimizing controls (as distinguished from relaxed, or measure-valued, controls). We ought to mention, however, that necessary conditions for relaxed minimizing controls are generally much easier to derive, and that we made a constant use of relaxed controls in proving the results that will be described below.

A detailed discussion and derivation of the results presented here as well as a number of illustrative examples are contained in references [4-12], some of which are as yet unpublished and one of which, [9], appears in print only in the Russian translation. The interested reader may obtain the preprints and the original of [9] from the author.

In order to introduce the concepts involved in these necessary conditions in as simple a context as possible, we shall begin by sketching some of these results as they apply to finite-dimensional nonlinear programming. If K is a convex subset of some euclidean space $\mathbb{R}^n$ and a continuous function $\varphi : K \to \mathbb{R}$ achieves a minimum at x_0 and has one-sided directional derivatives at x_0 then

$$D\varphi(x_0; x - x_0) \geq 0 \quad (x \in K) , \tag{1}$$

where

$$D\varphi(x_0; x - x_0) = \lim_{\alpha \to 0+} \frac{1}{\alpha} [\varphi(x_0 + \alpha[x - x_0]) - \varphi(x_0)] .$$

If in addition, for each $x \in K$, the function $\alpha \to \varphi(x_0 + \alpha[x - x_0]) : [0,1] \to \mathbb{R}$ has a second (one-sided) derivative $D^2\varphi(x_0; x - x_0)$ at $\alpha = 0$ then

$$\begin{aligned} &\text{either} \quad D\varphi(x_0; x - x_0) > 0 \\ &\text{or} \quad D^2\varphi(x_0; x - x_0) \geq 0 \quad (x \in K) . \end{aligned} \tag{2}$$

If φ does not have directional derivatives at x_0 then a different approach is necessary. One such approach, based on a generalization of the concept of a subgradient of a convex function, has been extensively studied by Clarke (see, e.g., [2,3] and other references listed in [3]). We shall consider, however, a different method that we investigated and applied to optimal control problems in [5-9]. This method is based on the concept of a "derivate container for φ" of which Clarke's "generalized derivative" is a special case. If we assume that φ can be extended to a Lipschitz continuous function defined over some neighborhood $\tilde{K}$ of K and if we determine any sequence (φ_i) of C^1 (continuously differentiable) functions with uniformly bounded derivatives such that

$$\lim_i \varphi_i = \varphi \quad \text{uniformly}$$

then we can construct a corresponding collection of sets $\Lambda\varphi(x)$ $(x \in \tilde{K})$ (which is a special form of a "derivate container for φ") by choosing, for all $\varepsilon > 0$, a positive integer $i^*(\varepsilon)$ and setting

$$\Lambda^{\varepsilon}\varphi(x) = \text{closure}\{\varphi_i'(v) \mid |v - x| < \varepsilon, i \geq i^*(\varepsilon), v \in \tilde{K}\}$$

(3) $$\Lambda\varphi(x) = \bigcap_{\varepsilon>0} \Lambda^{\varepsilon}\varphi(x) .$$

Each $\Lambda^{\varepsilon}\varphi$ and $\Lambda\varphi(x)$ is then a nonempty compact subset of $L(\mathbb{R}^n,\mathbb{R})$ and, if x_0 minimizes φ on K, we can make an assertion analogous to (1); namely

(1') $$\text{there exists } M \in \Lambda\varphi(x_0) \text{ such that } M(x - x_0) \geq 0 \quad (x \in K) .$$

In order to provide a degree of generality to statement (1') - and to analogous statements derived for more complicated optimization problems - we must indicate some general procedures for constructing derivate containers. One such procedure leads to the following construction [8, p. 16] that applies to any Lipschitz continuous function $\psi : A \to \mathbb{R}^m$, where A is an open subset of $\mathbb{R}^n$: By a known theorem of Rademacher, ψ is differentiable a.e. in A. If we define a derivate container for ψ exactly as we did for scalar-valued φ, then the sets

$$\partial^{\varepsilon}\psi(x) = \text{convex closure}\{\varphi'(v) \mid |v - x| < 2\varepsilon, \psi'(v) \text{ exists}\} \qquad (\varepsilon > 0),$$

$$\partial\psi(x) = \bigcap_{\varepsilon>0} \partial^{\varepsilon}\psi(x)$$

can be shown to satisfy this definition with $\partial^{\varepsilon}, \partial$ replacing $\Lambda^{\varepsilon}, \Lambda$. Furthermore, the special derivate container $\partial\psi(x)$ coincides with Clarke's "generalized derivative." We can also obtain [8, Remark 2.13, pp. 17-18] derivate containers $\Lambda\psi(x)$ that are sometimes "better" than $\partial\psi(x)$ (i.e., $\Lambda\psi(x)$ is a proper subset of $\partial\psi(x)$) by a "chain rule" construction: if $\psi = \psi_1 \circ \cdots \circ \psi_k$ and $\Lambda\psi_i(x)$ is a derivate container for ψ_i then

$$\Lambda\psi(x) = \{M_1 \cdot \cdots \cdot M_k \mid M_i \in \Lambda\psi_i(x)\}$$

is a derivate container for ψ ; in particular

$$\Lambda\psi(x) = \{M_1 \cdot \cdots \cdot M_k \mid M_i \in \partial\psi_i(x)\}$$

is always a derivate container for ψ.

If φ is C^1 then (1') implies (1) because then $\partial\varphi(x) = \{\varphi'(x)\}$. In general, assertions involving $\Lambda\varphi(x)$ for a Lipschitzian ψ are analogous to assertions involving the derivative of a C^1 function, the statement " $\psi'(x)$ has property P " being usually replaced by the statement " $\Lambda\psi(x)$ contains an element M that has property P."

Relations generalizing (1), (2) and (1') can be established if we wish to minimize φ^0 on K subject to the restriction $\varphi^1(x) = 0$, where φ^0 and φ^1 are continuous functions over K, or some neighborhood of K, into $\mathbb{R}$, respectively $\mathbb{R}^m$. By introducing a Lagrange multiplier (vector) ℓ, we can replace (1), respectively (1'), by

(1a) there exists $\ell = (\ell_0, \ell_1) \in \mathbb{R} \times \mathbb{R}^m$ such that
$$\ell \neq 0 , \quad \ell_0 \geq 0 , \quad \ell^T D(\varphi^0,\varphi^1)(x_0; x - x_0) \geq 0 \quad (x \in K) ,$$

respectively,

(1'a) if $\Lambda(\varphi^0,\varphi^1)(x_0)$ is a derivate container for (φ^0,φ^1) at x_0 then there exist $\ell = (\ell_0,\ell_1) \in \mathbb{R} \times \mathbb{R}^m$ and $M \in \Lambda(\varphi^0,\varphi^1)(x_0)$ such that
$$\ell \neq 0 , \quad \ell_0 \geq 0 , \quad \ell^T M(x - x_0) \geq 0 \quad (x \in K) .$$

The second order condition (2) generalizes in a more complicated manner described in Theorem 3.1.

We reach a higher degree of generality when we abandon the finite-dimensional framework and consider a convex subset K of a normed vector space $\mathcal{X}$, a Banach space $\mathcal{Y}$, a convex body $C \subset \mathcal{Y}$ and continuous functions

$$\varphi^0 : K \to \mathbb{R} , \quad \varphi^1 : K \to \mathbb{R}^m , \quad \Phi : K \to \mathcal{Y} .$$

These data define the optimization problem of minimizing φ^0 on $\{x \in K \mid \varphi^1(x) = 0, \Phi(x) \in C\}$, a problem which can serve as a framework for the study of relaxed control problems with unilateral restrictions. Here K can be interpreted as an appropriate collection of relaxed controls and φ^0, φ^1 and Φ as functions of the solution of the differential or functional-integral equation evaluated for the control x. We have derived a condition of order 0 for this problem (Theorem 2.6) in terms of a "directional derivate container" which is effectively a generalization of the ordinary derivate container to functions between infinite-dimensional spaces. We were also able to derive a condition of order 2 for problems of this kind that do not involve the "unilateral" or "functional" restriction $\Phi(x) \in C$.

The optimal control results presented below do not fit exactly within this framework because they refer to ordinary controls to which the optimization framework above does not always directly apply (because ordinary controls don't always form a set that is compact and convex with appropriate linear and topological structures). However, more complicated arguments yield a necessary condition of order 0 for ordinary controls (Theorem 2.3). We also present two different conditions of order 2 (Theorems 3.2 and 3.3), both of them obtained by applying Theorem 3.1 in two different contexts. Theorem 3.2 is obtained by embedding ordinary controls among relaxed controls. Theorem 3.3 is derived for more special problems satisfying stronger differentiability conditions and involving ordinary controls $u : [t_0,t_1] \to U$ with $u(t) \in U^{\#}(t)$ a.e., where U is finite-dimensional and each $U^{\#}(t)$ convex.

2. NECESSARY CONDITIONS OF ORDER 0

We shall state two theorems, one of them pertaining to the optimal control of ordinary differential equations with unilateral restrictions and the other to a general optimization problem.

Let $n,m,m_2 \in \{1,2,\ldots\}$, V be an open subset of $\mathbb{R}^n$, $T = [t_0,t_1] \subset \mathbb{R}$, T^h be a compact subset of T , A_0 and A_1

closed convex subsets of $\mathbb{R}^n$ and $\mathbb{R}^m$, A a convex body in $\mathbb{R}^{m_2}$ (i.e., a closed convex set with a nonempty interior), U a compact metric space, and $U^{\#}(t)$ $(t \in T)$ closed subsets of U such that the set $\{t \in T \mid U^{\#}(t) \cap G \neq \phi\}$ is measurable for every open $G \subset U$. We assume that, for all $(\tau,v,u) \in T \times V \times U$ and $t \in T^h$, $f(\cdot,v,u)$ is measurable, $f(\tau,\cdot,\cdot)$ and h^2 continuous and, for every compact $V^* \subset V$, the restrictions of

$$f(\tau,\cdot,u)\ ,\ h^0\ ,\ h^1\ ,\ h^2(t,\cdot) \qquad (\tau \in T, u \in U, t \in T^h)$$

to V^* have a common bound and a common Lipschitz constant.

We shall say that a couple (u,a_0) is *admissible* if $a_0 \in A_0$, $u : T \to U$ is measurable, the equation

$$y(t) = a_0 + \int_{t_0}^{t} f(\tau,y(\tau),u(\tau))d\tau \qquad (t \in T)$$

has a unique solution $y(u,a_0)$, and we have

$$u(t) \in U^{\#}(t) \text{ a.e.}, \quad h^1(y(u,a_0)(t_1)) \in A_1\ ,$$

$$h^2(t,y(u,a_0)(t)) \in A \qquad (t \in T^h)\ .$$

We shall say that $(\bar{u},\bar{a}_0)$ is a *minimizing couple* if $(\bar{u},\bar{a}_0)$ minimizes $h^0(y(u,a_0)(t_1))$ among all admissible couples (u,a_0).

We write $L(E,F)$ for the collection of linear operators between real vector spaces E and F; thus $L(\mathbb{R}^n,\mathbb{R}^m)$ represents the set of real $m \times n$ matrices. We write $C(K,Z)$ for the Banach space of bounded continuous functions on K to Z, endow each $\mathbb{R}^k$ with the euclidean norm, use the superscript ^{T} to denote a row vector or the transpose of a matrix, and write $\text{co}A$, $\overline{\text{co}}A$ and A^0 for the convex hull, convex closure and interior of A . We also denote by $\mathcal{Y}^*$ the topological dual of a real Banach space $\mathcal{Y}$ and by $M|_K$ the restriction of a function M to K .

DEFINITION 2.1 *Derivate container.* Let $j \in \{1,2,...\}$ and $\psi : T \times V \times U \to \mathbb{R}^j$ be such that, for $(t,v,u) \in T \times V \times U$, $\psi(\cdot,v,u)$ is measurable, $\psi(t,\cdot,\cdot)$ continuous and $\psi(t,\cdot,u)$ has a Lipschitz constant independent of t and u . A bounded collection

$$\{\Lambda^{\varepsilon}\psi(t,v,u) \mid \varepsilon > 0 \text{ , } (t,v,u) \in T \times V \times U\}$$

of nonempty subsets of $L(\mathbb{R}^n,\mathbb{R}^j)$, also denoted by $\Lambda^{\varepsilon}\psi$, is a *derivate container for* ψ *(with respect to* v *)* if

$$\Lambda^{\varepsilon}\psi(t,v,u) \subset \Lambda^{\varepsilon'}\psi(t,v,u) \qquad (\varepsilon' > \varepsilon)$$

and for every compact $V^* \subset V$ there exists a neighborhood $\tilde{V}$ of V^* in V and a sequence $\psi^p : T \times \tilde{V} \times U \to \mathbb{R}^j$ $(p = 1,2,...)$ of functions such that each ψ^p has a partial derivative ψ^p_v , both ψ^p and ψ^p_v are measurable in t and continuous in (v,u) , $\lim_p \psi^p = \psi$ uniformly on $T \times V^* \times U$, and for every $\varepsilon > 0$ there exist $i^* = i(\varepsilon,V^*)$ and $\delta^* = \delta(\varepsilon,V^*) > 0$ such that

$$\psi^p_v(t,v,u) \in \Lambda^{\varepsilon}\psi(t,w,u)$$

$$(p \geq i^* \text{ , } (t,w,u) \in T \times V^* \times U \text{ , } |v - w| \leq \delta^*) \text{ .}$$

We write $\Lambda\psi(t,v,u) = \bigcap_{\varepsilon>0} \Lambda^{\varepsilon}\psi(t,v,u)$.

THEOREM 2.2 ([8, Thm. 2.15, p. 19]) Let ψ be as described in Definition 2.1 and let, for $(t,w,u) \in T \times V \times U$,

$$\partial^{\varepsilon}\psi(t,w,u) = \overline{\text{co}}\{\psi_v(t,v,u) \;\Big|\; v \in V, |v - w| \le \varepsilon, \psi_v(t,v,u) \text{ exists}\} .$$

Then $\partial^{\varepsilon}\psi$ is a derivate container for ψ (with respect to v).

If $V_0, V_1, \ldots, V_k$ are open subsets of $\mathbb{R}^{j_0}, \ldots, \mathbb{R}^{j_k}$, the functions $\psi_i : T \times V_i \times U \to V_{i-1}$ $(i = 1, \ldots, k)$ satisfy similar conditions as ψ, $\Lambda^{\varepsilon}\psi_i$ is a derivate container for ψ_i, and

$$\psi(t,\cdot,u) = \psi_1(t,\cdot,u) \circ \cdots \circ \psi_k(t,\cdot,u) \qquad [(t,u) \in T \times U] ,$$

then

$$\Lambda^{\varepsilon}\psi(t,v,u) = \{M_1 \cdot \cdots \cdot M_k \mid M_i \in \Lambda^{\varepsilon}\psi_i(t,v,u)\}$$

defines a derivate container for ψ.

We define derivate containers for (h^0,h^1) and h^2 by treating each of them as a function on $T \times V \times U$.

THEOREM 2.3 ([9, XI.4.10]) Let $(\bar{u},\bar{a}_0)$ be a minimizing couple, $\bar{y} = y(\bar{u},\bar{a}_0)$, and let $\Lambda^{\varepsilon}f$, $\Lambda^{\varepsilon}(h^0,h^1)$ and $\Lambda^{\varepsilon}h^2$ be derivate containers for f, (h^0,h^1) and h^2. Then there exist $\ell = (\ell_0,\ell_1) \in \mathbb{R} \times \mathbb{R}^m$, $(H_0,H_1) \in \Lambda(h^0,h^1)(\bar{y}(t_1))$, $H_2 : T^h \to L(\mathbb{R}^n,\mathbb{R})$, $F : T \to L(\mathbb{R}^n,\mathbb{R}^n)$ and a nonnegative Radon measure ω on T^h such that H_2 is bounded and Borel measurable, F is (Legesgue) measurable,

(1) $\ell_0 \ge 0$, $\ell_0 + |\ell_1| + \omega(T^h) > 0$,

(2) $\omega(\{t \in T^h \mid h^2(t,\bar{y}(t)) \in A^o\}) = 0$,

$$(3)\quad H_2(t) \in \overline{co}\{M_1M_2 \mid M_1h^2(t,\bar{y}(t)) = \operatorname*{Max}_{a\in A} M_1a\ ,\ |M_1| \geq 1\ ,$$

$$M_2 \in \bigcap_{\varepsilon>0} \overline{co} \bigcup_{|\tau-t|\leq\varepsilon} \Lambda^{\varepsilon}h^2(\tau,\bar{y}(\tau))\}\quad \omega\text{ - a.e.},$$

$$(4)\quad F(t) \in co\ \Lambda f(t,\bar{y}(t),\bar{u}(t)) \quad \text{a.e. in } T\ ,$$

$$(5)\quad k(t)^T f(t,\bar{y}(t),\bar{u}(t)) = \operatorname*{Min}_{u\in U^{\#}(t)} k(t)^T f(t,\bar{y}(t),u)$$

$$\text{a.e. in } T\ ,$$

$$(6)\quad k(t_0)^T\bar{a}_0 = \operatorname*{Min}_{a_0\in A_0} k(t_0)a_0\ ,\quad \ell_1^T h^1(\bar{y}(t_1)) = \operatorname*{Max}_{a_1\in A_1} \ell_1^T a_1\ ,$$

where

$$k(t)^T = [\ell_0 H_0 + \ell_1^T H_1 + \int_{[t,t_1]} H_2(\tau)Z(\tau)^{-1}\omega(d\tau)]Z(t)\ ,$$

$$Z(t) = I + \int_t^{t_1} Z(\tau)F(\tau)d\tau \qquad (t \in T)$$

and I is the unit $n \times n$ matrix.

We observe that, in the special case where h_v^0, h_v^1, h_v^2 and $f_v(t,\cdot,\cdot)$ exist and are continuous, Theorem 2.3 yields the "usual" version of Pontryagin's principle and of the "transversality" conditons for unilateral problems (see, e.g., [4, VI.2.3, p. 357]).

We next consider the abstract optimization problem mentioned in Section 1. Let $\mathcal{X}$ be a real normed vector space, $\mathcal{Y}$ a real Banach space, K a convex and compact subset of $\mathcal{X}$, C a convex body in $\mathcal{Y}$, $m \in \{1,2,\ldots\}$ and $(\varphi^0,\varphi^1,\Phi) : K \to \mathbb{R} \times \mathbb{R}^m \times \mathcal{Y}$ a continuous function. If E and F are normed vector spaces, A a convex subset of E, $A^o \neq \phi$ and $\psi : A \to F$, we say that $\psi'(a)$ is the derivative of ψ at $a \in A$ if $\psi'(a) \in L(E,F)$, $\psi'(a) : E \to F$ is continuous and

$$\lim |x - a|^{-1}[\psi(x) - \psi(a) - \psi'(a)(x - a)] = 0$$

$$\text{as } x \to a \ , \ x \in A \ , \ x \neq a \ .$$

DEFINITION 2.5 *Directional derivate container.*[1] Let $\varphi = (\varphi^0, \varphi^1, \Phi)$. A collection $\{\Lambda^\varepsilon \varphi(x_0) \mid \varepsilon > 0\}$ of nonempty subsets of $L(X, \mathbb{R} \times \mathbb{R}^m \times Y)$ is a *directional derivate container for* $(\varphi^0, \varphi^1, \Phi)$ *at* x_0 if there exist continuous $\varphi_i : K \to \mathbb{R} \times \mathbb{R}^m \times Y$ $(i = 1,2,\ldots)$ such that

(1) $\lim_i \varphi_i = \varphi$ uniformly,

(2) $\Lambda^\varepsilon \varphi(x_0) \subset \Lambda^{\varepsilon'} \varphi(x_0)$ $\quad (\varepsilon' > \varepsilon)$,

(3) for every $\varepsilon > 0$, the set $\{M|_K \mid M \in \Lambda^\varepsilon \varphi(x_0)\}$

is an equicontinuous subset of $C(K, \mathbb{R}^m \times Y)$ and for every sequence (β_j) decreasing to 0 and every sequence (M_j) with $M_j \in \Lambda^{\beta_j} \varphi(x_0)$ there exist $J \subset (1,2,\ldots)$ and $M \in \Lambda \varphi(x_0)$ such that $\lim_j M_j x = Mx \quad (x \in K)$,

(4) for every $N \in \{1,2,\ldots\}$, $(x^1,\ldots,x^N) \subset K^N$ and $\varepsilon > 0$ there exist $\delta > 0$, $i^* \in \{1,2,\ldots\}$ and corresponding sets

$$T_N = \{(\omega^1,\ldots,\omega^N) \in \mathbb{R}^N \mid \omega^j \geq 0 \ , \ \sum_{j=1}^N \omega^j \leq 1\}$$

and

$$K^* = \{x_0 + \sum_{j=1}^N \omega^j (x^j - x_0) \mid (\omega^1,\ldots,\omega^N) \in \delta T_N\}$$

such that the functions

---

[1] *Definition 2.5 and Theorem 2.6 appear in a more general form in* [11].

$$(\omega^1,\ldots,\omega^N) \to \varphi(x_0 + \sum_{j=1}^{N} \omega^j(x^j - x_0)) : \delta T_N \to \mathbb{R} \times \mathbb{R}^m \times \mathcal{Y}$$

$$(i = 1,2,\ldots)$$

are continuously differentiable and for every $x' \in K^*$ and $i \geq i^*$ there exists $M \in \Lambda\varphi(x_0)$ satisfying

$$D\varphi_i(x';x - x') = M(x - x') \qquad (x \in K) .$$

We set $\Lambda\varphi(x_0) = \bigcap_{\varepsilon>0} \Lambda^\varepsilon\varphi(x_0)$ and observe that, by (2) and (3), $\Lambda\varphi(x_0) \neq \phi$.

THEOREM 2.6 [11] Let x_0 yield the minimum of φ^0 on the set $\{x \in K \mid \varphi^1(x) = 0 , \Phi(x) \in C\}$, and let $\Lambda^\varepsilon(\varphi^0,\varphi^1,\Phi)(x_0)$ be a directional derivate container for $(\varphi^0,\varphi^1,\Phi)$ at x_0 . Then there exist $\ell = (\ell_0,\ell_1,\ell_2) \in \mathbb{R} \times \mathbb{R}^m \times \mathcal{Y}^*$ and $M \in \Lambda(\varphi^0,\varphi^1,\Phi)(x_0)$ such that

$$\ell \neq 0 , \ \ell_0 \geq 0 , \ \ell \circ Mx_0 = \operatorname*{Min}_{x\in K} \ell \circ Mx , \ \ell_2\Phi(x_0) = \operatorname*{Max}_{c\in C} \ell_2 c .$$

We shall investigate elsewhere an application of Theorem 2.6 to the optimal relaxed control of functional-integral equations.

3. NECESSARY CONDITIONS OF ORDER 2

The necessary conditons of order 2 for optimal control problems will be stated in Theorems 3.2 and 3.3. The latter are based on the following general optimization test in which we use the notation

$$D\varphi(\bar{x};x - \bar{x}) = \bar{\varphi}'(0) , \quad D^2\varphi(\bar{x};x - \bar{x}) = \bar{\varphi}''(0) ,$$

where $\bar{\varphi}(\alpha) = \varphi(\bar{x} + \alpha[x - \bar{x}])$ is assumed to be defined for sufficiently small $|\alpha|$.

THEOREM 3.1 [10, Theorem 2.3] Let K be a convex subset of a vector space X, $D \subset X$, $\bar{x} \in K \cap D$, $m \in \{1,2,\ldots\}$ and $\varphi = (\varphi^0,\varphi^1) : D \to \mathbb{R} \times \mathbb{R}^m$. Assume that for every choice of $\xi = (x_0,x_1,\ldots,x_{m+1}) \in K^{m+2}$ there exists $\alpha_\xi > 0$ such that

$$\bar{x} + \sum_{j=0}^{m+1} a^j(x_j - \bar{x}) \in D$$

whenever $a = (a^0,\ldots,a^{m+1}) \in S(0,\alpha_\xi)$ (the euclidean open ball with center 0 and radius α_ξ) and that the function

$$a \to \varphi(\bar{x} + \sum_{j=0}^{m+1} a^j(x_j - \bar{x})) : S(0,\alpha_\xi) \to \mathbb{R} \times \mathbb{R}^m$$

has continuous second derivatives. If $\bar{x}$ minimizes φ^0 on the set $\{x \in K \cap D \mid \varphi^1(x) = 0\}$ then

(1) there exists $\ell = (\ell_0,\ell_1) \in \mathbb{R} \times \mathbb{R}^m$ such that
$$\ell \neq 0, \quad \ell_0 \in \{0,1\}, \quad \ell^T D\varphi(\bar{x};x - \bar{x}) \geq 0 \quad (x \in K).$$

Furthermore, if ℓ satisfies (1), $\ell_0 = 1$ and

$$K^* = \{x \in K \mid \ell^T D\varphi(\bar{x};x - \bar{x}) = 0\}$$

then either

(2a) there exists $\lambda \in \mathbb{R}^m$ such that
$$\lambda \neq 0, \quad \lambda^T D\varphi^1(\bar{x};x - \bar{x}) \geq 0 \quad (x \in K^*)$$

or

(2b) the set $\{D\varphi^1(\bar{x};x - \bar{x}) \mid x \in K^*\}$ contains the origin of $\mathbb{R}^m$ in its interior and, for every $\tilde{x} \in K^*$ with $D\varphi^1(\bar{x};\tilde{x} - \bar{x}) = 0$, we have

$$\ell^T D^2\varphi(\bar{x};\tilde{x} - \bar{x}) \geq 0 .$$

We next consider a less general optimal control problem than the one discussed in Section 2. Specifically, we neglect the unilateral restriction $h^2(t,y(t)) \in A \quad (t \in T^h)$, replace

$$y(t_0) \in A_0 , \ h^1(y(t_1)) \in A_1 \ \text{ by } \ y(t_0) = 0 , \ h^1(y(t_1)) = 0 ,$$

and replace the assumption of Lipschitz continuity by the assumption that h^1_v , h^1_{vv} , $f_v(t,\cdot,\cdot)$ and $f_{vv}(t,\cdot,\cdot)$ exist, have a common bound, and are continuous. Here the subscripts v and vv denote the first and second order partial derivatives with respect to the variable in V . While the elimination of the unilateral restriction and the assumption of second order differentiability severely detract from the generality of the problem, the old endpoint restrictions can be reduced to the new ones by an easy transformation [10, Section 2].

We refer to a measurable $u : T \to U$ as an *admissible control* if the equation

$$y(t) = \int_{t_0}^{t} f(\tau,y(\tau),u(\tau))d\tau \qquad (t \in T)$$

has a unique solution $y(u)$ and we have

$$u(t) \in U^{\#}(t) \ \text{ a.e.}, \ \ h^1(y(u)(t_1)) = 0 .$$

We call and admissible control a *minimizing control* if it yields the minimum of $h^0(y(u)(t_1))$ among admissible controls u . An admissible control $\bar{u}$ is *extremal* if it satisfies Pontryagin's "minimum" principle, i.e., there exists $\ell = (\ell_0,\ell_1) \in \mathbb{R} \times \mathbb{R}^m$

such that $\ell \neq 0$, $\ell_0 = 0$ or 1 , and

$$z(t)^T f(t,y(\bar{u})(t),\bar{u}(t)) = \min_{u \in U^{\#}(t)} z(t)^T f(t,y(\bar{u})(t),u)$$

a.e. in T ,

where

$$z(t)^T = \sum_{i=0}^{1} \ell_i^T h_v^i(y(\bar{u})(t_1))Z(t) ,$$

Z is the solution of the matrix-differential equation

$$Z(t) = I + \int_t^{t_1} Z(\tau)f_v(\tau,y(\bar{u})(\tau),\bar{u}(\tau))d\tau \quad (t \in T)$$

and I is the unit $n \times n$ matrix. The function z is the "dual" function of Pontryagin's principle.

For an (admissible) extremal control $\bar{u}$ and a corresponding vector ℓ and functions Z and z , we set

$$\bar{y} = y(\bar{u}) , \quad g(v) = \sum_{i=0}^{1} \ell_i^T h^i(v) ,$$

$$H(t,v,u) = z(t)^T f(t,v,u) ,$$

$$M(s) = g_{vv}(\bar{y}(t_1)) + \int_s^{t_1} (Z(\tau)^T)^{-1} H_{vv}(\tau,\bar{y}(\tau),\bar{u}(\tau))Z(\tau)^{-1} d\tau ,$$

$$U'(t) = \{u' \in U^{\#}(t) \mid H(t,\bar{y}(t),u') = H(t,\bar{y}(t),\bar{u}(t))\} ,$$

$$\mathcal{U}^* = \{u : T \to U \mid u \text{ is measurable, } u(t) \in U'(t) \text{ a.e.}\} ,$$

$$\tilde{f}(u)(t) = Z(t)[f(t,\bar{y}(t),u(t)) - f(t,\bar{y}(t),\bar{u}(t))] ,$$

$$\tilde{H}_v(u)(t) = H_v(t,\bar{y}(t),u(t)) - H_v(t,\bar{y}(t),\bar{u}(t)) ,$$

where M , g_{vv} and H_{vv} are treated as square matrices.

THEOREM 3.2 [10, Thm. 2.2] Let $\bar{u}$ be a minimizing control. Then $\bar{u}$ is extremal. If the corresponding vector $\ell = (\ell_0, \ell_1)$ is such that $\ell_0 = 1$ then either

(a) there exists $\lambda \in \mathbb{R}^m$ such that $\lambda \neq 0$ and
$\lambda^T h_v^1(\bar{y}(t_1))\tilde{f}(u)(t) \geq 0$ a.e. in T $(u \in \mathcal{U}^*)$

or

(b) the set

$$\operatorname{co}\{h_v^1(\bar{y}(t_1)) \int_{t_0}^{t_1} \tilde{f}(u)(\tau)d\tau \mid u \in \mathcal{U}^*\}$$

contains 0 in its interior and, for every choice of measurable $(\hat{f},\hat{H}) : T \to \mathbb{R}^n \times L(\mathbb{R}^n,\mathbb{R})$ such that

$$(\hat{f}(t),\hat{H}(t)) \in \operatorname{co}\{(\tilde{f}(u)(t),\tilde{H}_v(u)(t)) \mid u \in \mathcal{U}^*\} \quad \text{a.e.}$$

$$h_v^1(\bar{y}(t_1)) \int_{t_0}^{t_1} \hat{f}(\tau)d\tau = 0 ,$$

we have

$$\int_{t_0}^{t_1} [\hat{f}(s_1)^T M(s_1) + \hat{H}(s_1)Z(s_1)^{-1}]ds_1 \int_{t_0}^{s_1} \hat{f}(s_2)ds_2 \geq 0 .$$

Finally,

(c) if $\ell_0 = 1$ and (b) is valid then there exists a set $T' \subset T$ whose complement is a null-set and such that

$$\sum_{i=0}^{k} \sum_{j=0}^{i} \beta_i\beta_j e(\tau_i - \tau_j)[\hat{f}_i^T M(\tau_i) + \hat{H}_i Z(\tau_i)^{-1}]\hat{f}_j \geq 0$$

(where $e(0) = \frac{1}{2}$ and $e(a) = 1$ for $a > 0$) provided

$$k \in \{m,m+1,\ldots\} , \quad \tau_i \in T' , \quad \beta_i > 0 ,$$

$$t_0 < \tau_0 \leq \tau_1 \leq \cdots \leq \tau_k \leq t_1 ,$$

$$(\hat{f}_i,\hat{H}_i) \in \text{co}\{(Z(\tau_i)f(\tau_i,\bar{y}(\tau_i),u) - a_i\ ,$$
$$H_v(\tau_i,\bar{y}(\tau_i),u) - b_i) \mid u \in U'(\tau_i)\}\ ,$$

$$a_i = Z(\tau_i)f(\tau_i,\bar{y}(\tau_i),\bar{u}(\tau_i))\ ,\quad b_i = H_v(\tau_i,\bar{y}(\tau_i),\bar{u}(\tau_i))\ ,$$

$$h_v^1(\bar{y}(t_1)) \sum_{j=0}^{k} \beta_j\hat{f}_j = 0\ ,$$

$$0 \in \text{interior co}\{h_v^1(\bar{y}(t_1))\hat{f}_j \mid j = 0,1,\ldots,k\}\ .$$

Finally, we state a second order condition which generalizes a classical test of the calculus of variations (the "accessory minimum problem" [1, p. 228]) to problems with restricted controls. We consider the same problem as in Theorem 3.2 but with the following additional assumptions: U is a bounded open subset of some $\mathbb{R}^k$; each $U^{\#}(t)$ is a closed convex subset of U ; each $f(\cdot,v,u)$ is measurable and each $f(t,\cdot,\cdot)$ has uniformly bounded continuous derivatives of first and second order which we denote by f_v , f_u , f_{vv} , f_{vu} and f_{uu} .

For an extremal control $\bar{u}$ and a corresponding vector ℓ , functions $\bar{y} = \bar{y}(u)$, Z , z , H , M and g , we write $\bar{X}(t)$ for the expression $X(t,\bar{y}(t),\bar{u}(t))$ (thus $\bar{H}_{vu}(t) = H_{vu}(t,\bar{y}(t),\bar{u}(t))$, $\overline{Zf}_u(t) = Z(t)f_u(t,\bar{y}(t),\bar{u}(t))$ etc.). We also set

$$\Omega^* = \{\omega : T \to \mathbb{R}^k \mid \omega \text{ is measurable, } \omega(t) \in U^{\#}(t) - \bar{u}(t)$$
$$\text{and } z(t)^T f_u(t,\bar{y}(t),\bar{u}(t))\omega(t) = 0 \text{ a.e.}\}\ .$$

THEOREM 3.3 [12] Let $\bar{u}$ be a minimizing control. Then $\bar{u}$ is extremal. If the corresponding vector $\ell = (\ell_0, \ell_1)$ is such that $\ell_0 = 1$ then either

(a) there exists $\lambda \in \mathbb{R}^m$ such that $\lambda \neq 0$ and

$$\lambda^T h_v^1(\bar{y}(t_1))\overline{Zf}_u(t)\omega(t) \geq 0 \text{ a.e. in } T \qquad (\omega \in \Omega^*)$$

or

(b) the set

$$\text{co}\{h_v^1(\bar{y}(t_1)) \int_{t_0}^{t_1} \overline{Zf}_u(\tau)\omega(\tau)d\tau \mid \omega \in \Omega^*\}$$

contains 0 in its interior and, for every choice of $\omega \in \Omega^*$ with

$$h_v^1(\bar{y}(t_1)) \int_{t_0}^{t_1} \overline{Zf}_u(\tau)\omega(\tau)d\tau = 0 ,$$

we have

$$\int_{t_0}^{t_1} \omega(s_1)^T\{[\overline{Zf}_u(s_1)^T M(s_1)$$

$$+ \bar{H}_{uv}(s_1)Z(s_1)^{-1}] \int_{t_0}^{s_1} \overline{Zf}_u(s_2)\omega(s_2)ds_2$$

$$+ \tfrac{1}{2}\bar{H}_{uu}(s_1)\omega(s_1)\}ds_1 \geq 0 .$$

REFERENCES

1. Bliss, G. A. "Lectures on the Calculus of Variations." The Univ. of Chicago Press, 1946.

2. Clarke, F. H. Generalized gradients and applications, *Trans. Amer. Math. Soc.* *205*(1975), pp. 247-267.

3. Clarke, F. H. Generalized gradients of Lipschitz functionals, MRC Technical Summary Report, Aug. 10, 1976, Mathematics Research Center, University of Wisconsin, Madison.

4. Warga, J. "Optimal Control of Differential and Functional Equations." Academic Press, New York, 1972.

5. Warga, J. Necessary conditions without differentiability assumptions in optimal control, *J. Diff. Eqs. 18*(1975), pp. 40-61.

6. Warga, J. Necessary conditions without differentiability assumptions in unilateral control problems, *J. Diff. Eqs. 21*(1976), pp. 25-38.

7. Warga, J. Controllability and necessary conditions in unilateral problems without differentiability assumptions, *SIAM J. Control Optim. 14*(1976), pp. 546-573.

8. Warga, J. Derivate containers, inverse functions and controllability, in "Calculus of Variations and Control Theory," D. L. Russell, ed., Academic Press, New York, 1976.

9. Warga, J. Chapter XI, appended to the Russian translation of "Optimal Control of Differential and Functional Equations," Nauka, Moscow, 1977.

10. Warga, J. A second order condition that strengthens Pontryagin's maximum principle, *J. Diff. Eqs.*, to appear.

11. Warga, J. Controllability and a multiplier rule for nondifferentiable optimization problems, preprint.

12. Warga, J. A second order "Lagrangian" condition for restricted control problems, *J. Optimization Theory Appl.*, to appear.

NUMERICAL METHODS FOR THE SOLUTION OF TIME OPTIMAL CONTROL PROBLEMS FOR HEREDITARY SYSTEMS

M. Q. Jacobs*
W. C. Pickel*

University of Missouri
Columbia, Missouri

A number of years ago Neustadt [1] described a method for synthesizing time optimal controllers for linear systems of ordinary differential equations. Computational results using the method and its variations have been discussed in a number of papers [e.g., 2, 3, 4, 5]. These methods were adapted to treat certain classes of linear hereditary systems in [6, 7, 8, 9, 10]. In this paper we describe how these methods can be adapted to handle classes of problems that cover quite general linear, autonomous, neutral functional differential equations.

Let $M_{p,q}$ be the collection of all real $p \times q$ matrices. We shall use $M_{p,1}$ and R^p as equivalent notations. Points x in R^p will be denoted by $[x^1,\ldots,x^p]^T$. Let $X : R \to M_{n,n}$ be measurable and essentially bounded on every compact interval in

* *This research was supported in part by National Science Foundation Grant NSF MCS77-02613.*

ISBN 0-12-632250-3

R. A matrix B in $M_{n,r}$ is given. A convolution operator K is defined by

(1) $$K(u)(t) = \int_{t_0}^{t} X(t-s)Bu(s)ds\ , \quad t_0 \leq t \leq t_1\ ,$$

where $u : [t_0,t_1] \to R^r$ is an L_1 function. Let U be a non-empty, compact, convex subset of R^r, and let a real number t_0 be given. A triple $\{u,t_0,t_1\}$ is called an *admissible control triple* if $t_0 \leq t_1 \in R$ and $u : [t_0,t_1] \to U$ is measurable. The function u in the admissible control triple is simply called an *admissible control* (or controller).

Let $z : R \to R^n$ be a continuous function. The *time optimal control problem* which we shall discuss is that of finding an admissible control triple $\{\bar{u},t_0,\bar{t}_1\}$ such that

(2) $$K(\bar{u})(\bar{t}_1) = z(\bar{t}_1)$$

and

(3) $$K(u)(t_1) \neq z(t_1)$$

for any admissible control triple $\{u,t_0,t_1\}$ with $t_1 < \bar{t}_1$. We say that $\bar{t}_1$ is a *local minimum time* if there is an admissible control triple $\{\bar{u},t_0,\bar{t}_1\}$ such that (2) is satisfied and there is a $\delta > 0$ such that (3) is satisfied for all admissible control triples $\{u,t_0,t_1\}$ with $\bar{t}_1 - \delta < t_1 < \bar{t}_1$.

As a motivating example for this type of time optimal control problem we consider a controlled system of linear neutral functional differential equations [11, p. 300ff],

(4) $$\frac{d}{dt}\mathcal{D}(x_t) = L(x_t) + Bu(t)\ ,$$

where $\mathcal{D}, L : C([-h,0],R^n) \to R^n$ have the form

(5) $$\mathcal{D}(\varphi) = \varphi(0) - \int_{-h}^{0} [d\lambda(s)]\varphi(s) ,$$

and

(6) $$L(\varphi) = \int_{-h}^{0} [d\sigma(s)]\varphi(s) , \quad h > 0 ,$$

where $\lambda,\sigma : [-h,0] \to M_{n,n}$ are of bounded variation and

$$\lim_{\varepsilon \to 0^+} \mathrm{Var}_{[-\varepsilon,0]}\lambda = 0 .$$

Here x_t has its usual meaning, cf. [11]. Let $\{u,t_0,t_1\}$ be an admissible control triple, and let $\varphi \in C([-h,0],R^n)$ be given. Then $x(t,t_0,\varphi,u)$ will be used to denote the unique solution of (4) on $[t_0,t_1]$ satisfying the initial condition

(7) $$x_{t_0} = \varphi .$$

We observe that $x(t,t_0,\varphi,u) = x(t,t_0,\varphi,0) + x(t,t_0,0,u)$ and that

(8) $$x(t,t_0,0,u) = \int_{t_0}^{t} X(t - s)Bu(s)ds , \quad t_0 \leq t \leq t_1 ,$$

where $X(t)$ is the $n \times n$ transition matrix for the system (4), i.e., $t \to X(t)$, $t \geqq -h$, is locally of bounded variation and satisfies

$$\mathcal{D}(X_t) = I + \int_{0}^{t} L(X_s)ds , \quad t \geqq 0 ,$$

(9) $$X(0) = I , \quad X(t) = 0 , \quad t < 0 ,$$

where I is the $n \times n$ identity matrix (see [11, Sec. 12.8]). If we define $z(t) = -x(t,t_0,\varphi,0)$ and put $K(u)(t) = x(t,t_0,0,u)$, then the problem of minimizing the time t_1 such that $x(t_1, t_0, \varphi, u) = 0$ using admissible control triples $\{u, t_0, t_1\}$ falls into the category of problems we introduced in (1), (2), and (3).

Returning to the original problem (1), (2), (3) we define

$$C(t) = \{x \in R^n : x = K(u)(t), \{u, t_0, t\} \text{ admissible}\}, \tag{10}$$

$t \geq t_0$. The sets $C(t)$, $t \geq t_0$, are nonempty, compact, and convex subsets of R^n. The support functional of a nonempty, compact, convex set $A \subset R^n$ is defined [12] by the equation

$$H(y|A) = \sup_{x \in A} \langle y, x \rangle, \quad y \in R^n, \tag{11}$$

where $\langle y, x \rangle = \Sigma y^i x^i$. Let K denote the collection of all nonempty, compact, and convex subsets of R^n. The set K is assumed to carry the topology induced by the Hausdorff metric. It can be shown [12] that a function $H : R^n \to R$ is the support functional of an element $A \in K$ if and only if

$$\begin{aligned} &\text{(i)} \quad H(y_1 + y_2) \leq H(y_1) + H(y_2) \qquad \textit{(subadditive)} \\ &\text{(ii)} \quad H(ty) = tH(y), \; t \geq 0 \qquad \textit{(positively homogeneous)} \end{aligned} \tag{12}$$

for all $y, y_i \in R^n$, $i = 1,2$. Let H denote the collection of all continuous and positively homogeneous functions $H : R^n \to R$ with norm defined by $\|H\| = \sup\{|H(y)| : \|y\| \leq 1\}$. Then H is a Banach space and K is isomorphic to a convex cone in H (see [12]).

One can verify [13] that

$$\begin{aligned} H(y|C(t)) &= \int_{t_0}^{t} H(B^T X(t-s)^T y | U) ds \\ &= \int_0^{t-t_0} H(B^T X(s)^T y | U) ds, \end{aligned} \tag{13}$$

where $C(t)$ is given in (10). Since X is essentially bounded one can verify that the functions $t \to H(y|C(t))$, $t_0 \leq t \leq b$, with y being an element of the closed unit ball in R^n are equi-Lipschitzian. Hence, it follows that $t \to C(t)$ is continuous in the Hausdorff metric on K.

It is perhaps worth noting that if $0 \in U$, then the sets $C(t)$, $t \geq t_0$, are nondecreasing, i.e., $t_0 \leq s < t$ implies $C(s) \subset C(t)$. This follows at once from (13) since $0 \in U$ implies $H(B^T X(s)^T y|U) \geq 0$, $s \geq 0$. Thus $t \to H(y|C(t))$ is nondecreasing. One can establish that if $\bar{t}_1$ is at least a local minimum time for (1), (2), (3), then $z(\bar{t}_1)$ lies on the boundary of $C(\bar{t}_1)$. Hence there is a *nonzero* $y \in R^n$ such that

$$\langle y, z(\bar{t}_1)\rangle = H(y|C(\bar{t}_1)), \tag{14}$$

and from (14) it follows that a local optimal control triple $\{\bar{u}, t_0, \bar{t}_1\}$ must satisfy (2) and

$$H(B^T X(\bar{t}_1 - s)^T y|U) = \langle B^T X(\bar{t}_1 - s)^T y, \bar{u}(s)\rangle \quad \text{a.e.} \quad s \in [t_0, \bar{t}_1]. \tag{15}$$

To solve the time optimal control problem one needs to determine how (15) should be strengthened in order to give sufficient conditions for an optimal controller. Some results of this type were given in [10]. What we shall describe here is an extension of Pickel's local synthesis theorem [8] which was given for systems of the form

$$\dot{x}(t) = Ax(t) + B_0 u(t) + B_1 u(t - h). \tag{16}$$

Systems of the form (16) are closely related to certain neutral equations of the form (4) (see [8, 10]). Define

$$g(t,y) = H(y|C(t)) - \langle y, z(t)\rangle, \quad t \geq t_0, \quad y \in R^n. \tag{17}$$

The function $t \to g(t,y)$, $t \geqq t_0$, $y \in R^n$ is said to be *locally increasing at* $(t_1,y_1) \in R \times R^n$, $t_1 \geqq t_0$, if there is a $\delta > 0$ and an open neighborhood N of y_1 such that $g(s,y) < g(t,y)$ for $y \in N$ and $t_1 - \delta < s < t < t_1 + \delta$. A nonzero $y \in R^n$ *essentially determines the extremal control* $\bar{u}$ in (15) on the interval $[t_0,t_1]$, and we write $\bar{u} = u(\cdot,t_1,y)$, if any other extremal control u satisfying (15) must have the property that $X(t_1 - s)B(u(s) - \bar{u}(s)) = 0$ a.e. on $[t_0,t_1]$. Define

$$(18) \qquad P(t) = \{y \in R^n : y \neq 0, g(t,y) = 0\} , \quad t \geqq t_0 .$$

One can establish the following local synthesis theorem by methods similar to those used by Pickel [8]. *If* $\bar{t}_1$ *is a local minimum time and* $t \to g(t,y)$ *is locally increasing at* $(\bar{t}_1,y_1)$, $y_1 \in P(\bar{t}_1)$, *then* $\bar{t}_1$ *and the vectors* $y \in P(\bar{t}_1)$ *maximize the time for which* $g(t,y) = 0$ *on some neighborhood of* $\{\bar{t}_1\} \times P(\bar{t}_1)$. *Conversely, if* $\bar{t}_1$ *maximizes the time for which* $g(t,y) = 0$ *on some neighborhood of* $\{\bar{t}_1\} \times P(\bar{t}_1)$, *and if* $t \to g(t,y)$ *is locally increasing at* $(\bar{t}_1,y_1)$, $y_1 \in P(\bar{t}_1)$, *then* $\bar{t}_1$ *is a local minimum time. Moreover, if there is a* $\bar{y} \in P(\bar{t}_1)$ *which essentially determines the extremal control* $\bar{u} = u(\cdot,\bar{t}_1,\bar{y})$, *then* $\{\bar{u},t_0,\bar{t}_1\}$ *is a local optimal control triple.*

The above principle essentially reduces the time optimal control problem to one of computing the local maxima of a function $y \to F(y)$ defined implicitly by $g(F(y),y) = 0$. Thus from a practical viewpoint we would like g to be a C^1 function so that F would be C^1 and we could calculate the zeros of ∇F by the method of steepest ascent or one of its variations. The smoothness of g can be analyzed by taking advantage of well-known properties of the support functional $H(y|C(t))$ in (13). That is, x is an exposed point of $C(t)$ if and only if there is a $y_1 \in R^n$, $y_1 \neq 0$, such that $y \to H(y|C(t))$ is differentiable at y_1 and $\nabla_y H(y_1|C(t)) = x$ (see [12]). Of course, $y \to H(y|C(t))$ is differentiable at every point in $R^n \backslash \{0\}$ if and only if $C(t)$ is strictly convex. One can show, cf. [8, 14], that x is an exposed point of $C(t)$ if and only if there is a nonzero $y \in R^n$ which essentially determines an extremal control $u(\cdot,t,y)$ and $x = K(u(\cdot,t,y))(t)$. Thus for such an exposed point $x = K(u(\cdot,t,y))(t)$ we have $\nabla_y g(t,y) = K(u(\cdot,t,y))(t) - z(t)$. Now we note that if z is at least absolutely continuous, then

$$\frac{\partial}{\partial t} g(t,y) = H(B^T X(t - t_0)^T y|U) - \langle y,\dot{z}(t)\rangle \tag{19}$$

for almost every t. Stronger assertions may be obtained if further regularity assumptions are imposed on X and z. In most of the examples X will usually be at least piecewise analytic, but not necessarily continuous, and z is at least differentiable. Thus, if the time derivative in (19) is continuous and positive at (t_1,y_1), then $t \to g(t,y)$ is locally increasing at (t_1,y_1).

We remark that if an admissible control u exists which satisfies the objective

$$K(u)(t_1) = z(t_1), \tag{20}$$

then the time optimal control problem has a solution. A variety of conditions can assure that such an admissible control exists. For example, suppose $t_0 = 0$,

$$\lim_{t\to\infty} z(t) = 0, \tag{21}$$

$0 \in \text{int}(U)$, and the controllability Gramian

$$G(t) = \int_0^t X(s)BB^T X(s)^T ds \tag{22}$$

has rank n for some positive t. Then an admissible control satisfying (20) does exist.

Let A_i, $i = -1,0,1$ be elements of $M_{n,n}$. Suppose that in (4) the operator $\mathcal{D}$ has the form $\mathcal{D}(\varphi) = \varphi(0) - A_{-1}\varphi(-h)$, and that L has the form $L(\varphi) = A_0\varphi(0) + A_1\varphi(-h)$. If the initial data is sufficiently smooth, then (4) can be written as

$$\dot{x}(t) = A_{-1}\dot{x}(t-h) + A_0x(t) + A_1x(t-h) + Bu(t) . \tag{23}$$

The transition matrix $X(t)$ in (9) for such a system is analytic on the intervals $(k-1)h < t < kh$, $k = 1,2,3,\ldots$ with at most simple jumps at kh, $k = 0,1,2,\ldots$. Let $t_0 = 0$, and let $U = K^r$, the collection of all u in R^r such that $|u^i| \leq 1$, $i = 1,\ldots,r$. As in our earlier discussion of (4) we let

$$z(t) = -x(t,0,\varphi,0) . \tag{24}$$

Then (21) will be satisfied for the z in (24) if the roots of the characteristic equation

$$\det(z(I - A_{-1}e^{-zh}) - A_0 - A_1e^{-zh}) = 0 \tag{25}$$

are bounded above by an $a < 0$ [11]. The controllability Gramian (22) where $X(t)$ is the transition matrix for (23) will have rank n for some positive t if at least one of the controllability matrices

$$[B, A_iB, \ldots, A_i^{n-1}B] , \quad i = -1,0 \tag{26}$$

has rank n [15].

Pickel [9] has written algorithms and implemented them with computer programs for solving numerically the time optimal control problem for a restricted class of systems of the form (23) which are related to systems of the form (16). We will not go into a discussion of these algorithms here, but we will discuss briefly the mathematical solution of two examples in two dimensional space.

EXAMPLE 1 Let $n = 2$. Suppose $r = 1$ and $U = K^1$. The neutral system considered is

$$\dot{x}(t) = A_{-1}\dot{x}(t - 1) + Bu(t) \tag{27}$$

where

$$A_{-1} = \begin{bmatrix} 0 & 0 \\ 1 & 0 \end{bmatrix}, \quad B = \begin{bmatrix} 1 \\ 0 \end{bmatrix}.$$

The initial condition for this problem is

$$x_0 = \varphi$$

where $\varphi : [-1,0] \to R^2$ is C^1 and the first coordinate φ^1 of φ is a constant function. We note that the controllability matrix $[B, A_{-1}B]$ has rank 2 and that the Gramian $G(t)$ has rank 2 for $t > 2$. However, the roots of (25) are 0. One can show that the collection D_0 of all C^1 functions $\varphi : [-1,0] \to R^2$ with φ^1 a constant function that can be steered to the origin in R^2 in finite time is precisely those functions which satisfy the additional constraint

$$|\varphi^1(0) - \varphi^2(0)| \leqq 1 .$$

The optimal time function $\varphi \to T(\varphi)$ for $\varphi \in D_0$ is given by

$$T(\varphi) = \begin{cases} 1 + |\varphi^2(0)|, & \varphi^2(0) \neq 0 \\ |\varphi^1(0)|, & \varphi^2(0) = 0 . \end{cases}$$

In this problem the optimal controller is generally not uniquely determined. However, if $\varphi^2(0) \geqq 1$ and $\varphi^2(0) = \varphi^1(0) + 1$, then the optimal control is unique, and if we let $T = T(\varphi)$, it is given by $\bar{u}(t) = -1$, $0 \leqq t \leqq T - 1$, and $\bar{u}(t) = +1$, $T - 1 < t \leqq T$. We omit the other cases. For more details consult [7, Example 7.4] and [8, Chap. III].

EXAMPLE 2 Again let $n = 2$, $r = 1$, $U = K^1$. The system is

$$\dot{x}(t) = A_{-1}\dot{x}(t - 1) + A_0 x(t) + Bu(t)$$

where

$$A_0 = \begin{bmatrix} 0 & 1 \\ 0 & 0 \end{bmatrix}, \quad A_{-1} = -A_0, \quad B = \begin{bmatrix} 0 \\ 1 \end{bmatrix}.$$

The initial condition is $x_0 = \varphi$ where $\varphi : [-1,0] \to R^2$ is C^1 and φ^2 is a constant function on $[-1,0]$. We denote the collection of all such C^1 functions by D_0. One can show [8] that any function in D_0 can be steered to the origin in R^2 in finite time using an admissible control. For the complete analysis of this problem we refer to [8]. We will indicate here only a typical situation. If φ is in the region S defined by

$$S = \{\varphi \in D_0 : \varphi^2(0) > -2, \varphi^1(0) \geqq -((\varphi^2(0))^2 + 2\varphi^2(0) - 6)/2\}$$
$$\cup \{\varphi \in D_0 : \varphi^2(0) \leqq -2, \varphi^1(0) \geqq ((\varphi^2(0))^2 - 2\varphi^2(0) - 2)/2\},$$

then the optimal time function $T = T(\varphi)$ is given by

$$T = \varphi^2(0) + \sqrt{2(\varphi^2(0))^2 + 4\varphi^2(0) + 4\varphi^1(0) + 4}.$$

The unique optimal control is

$$\bar{u}(t) = \begin{cases} \operatorname{sgn}(T - t - s - 1) , & 0 \leqq t < T - 1 \\ \operatorname{sgn}(T - t - s) , & T - 1 \leqq t \leqq T , \end{cases}$$

where $s = s(\varphi) = 1 + (\varphi^2(0) - T(\varphi))/2$.

As a final remark we note that the time optimal control problem for neutral systems of the form

$$\dot{x}^1(t) = \dot{x}^2(t-h) + a_0 x^1(t) + b_0 x^2(t) + a_1 x^1(t-h) + b_1 x^2(t-h)$$

$$\dot{x}^2(t) = -a_1 x^1(t) - b_1 x^2(t) + u(t)$$

can be completely analyzed by the methods in [8]. The controllability Gramian $G(t)$ for such a system always has rank 2 if $t > 2h$. Moreover, since the roots of the characteristic equation (25) are in this case precisely the eigenvalues of the matrix

$$A = \begin{bmatrix} a_0 & b_0 \\ -a_1 & -b_1 \end{bmatrix} ,$$

one can readily check whether (21) is satisfied. Systems of the above type have been discussed in some detail in [8, 10].

REFERENCES

1. Neustadt, L. W. Synthesizing time optimal control systems, *J. Math. Anal. Appl.* *1*(1960), pp. 484-493.

2. Fadden, E. J. and E. G. Gilbert. Computational aspects of the time optimal control problem, *in* "Computing Methods in Optimization Problems." (A. V. Balakrishnan, L. W. Neustadt, eds.), pp. 167-192. Academic Press, New York, 1964.

3. Gilbert, E. G. The application of hybrid computers to the iterative solution of optimal control problems, *in* "Computing Methods in Optimization Problems." (A. V. Balakrishnan, L. W. Neustadt, eds.), pp. 261-284. Academic Press, New York, 1964.

4. Paiewonsky, B., P. Woodrow, W. Brunner and P. Halbert. Synthesis of optimal controllers using hybrid analog-digital computers, *in* "Computing Methods in Optimization Problems." (A. V. Balakrishnan, L. W. Neustadt, eds.), pp. 285-303. Academic Press, New York, 1964.

5. Paiewonsky, B. Synthesis of optimal controls, *in* "Topics in Optimization." (G. Leitmann, ed.), pp. 391-416. Academic Press, New York, 1967.

6. Oguztoreli, M. N. "Time-Lag Control Systems." Academic Press, New York, 1966. [Chap. 10]

7. Banks, H. T., M. Q. Jacobs and M. R. Latina. The synthesis of optimal controls for linear time optimal problems with retarded controls, *JOTA 8*(1971), pp. 319-366.

8. Pickel, W. C. Numerical Methods for Time Optimal Control of Linear Systems with Applications to Neutral Systems, Doctoral Dissertation, University of Missouri - Columbia, 1977.

9. Pickel, W. C. Algorithms for the solution of time optimal control problems involving hereditary systems, University of Missouri - Columbia, Tech. Report, 1976.

10. Jacobs, M. Q. and W. C. Pickel. Time optimal control of hereditary systems, *in* "Proceedings of the 14th Annual Allerton Conference on Circuit and System Theory." pp. 528-534. University of Illinois, Urbana-Champaign, 1976.

11. Hale, J. K. "Theory of Functional Differential Equations." Second Ed. Springer-Verlag, New York, 1977.

12. Rockafellar, R. T. "Convex Analysis." Princeton University Press, Princeton, N.J., 1970.

13. Goodman, G. S. Support functions and the integration of set-valued mappings, *in* Vol. II of "Control Theory and Topics in Functional Analysis." pp. 281-296. International Atomic Energy Agency, Vienna, 1976.

14. Hermes, H. and J. P. LaSalle. "Functional Analysis and Time Optimal Control." Academic Press, New York, 1969.

15. Jacobs, M. Q. and C. E. Langenhop. Criteria for function space controllability of linear neutral systems, *SIAM J. Control and Opt.* *14*(1976), pp. 1009-1048.

SOLVING OPTIMIZATION PROBLEMS

Magnus R. Hestenes

University of California, Los Angeles
Los Angeles, California

One of the important aspects of optimization theory is that of finding the numerical solution of an optimization problem. The purpose of this paper is to discuss several direct methods for obtaining a numerical solution of such a problem. We pay particular attention to conjugate direction and multiplier methods. These methods are closely related to gradient methods and Newton's method for minimizing a real valued function.

Conjugate direction methods were introduced by E. Stiefel and the author. Extensions have been given by several authors and, in particular, by Davidon, Fletcher and Powell. A comprehensive treatment of these and related methods will be given in a forthcoming book by the author entitled "Conjugate Direction Methods in Optimization." References to the literature on the subject matter found in this paper can be found in the recent

ISBN 0-12-632250-3

book by the author entitled "Optimization Theory, The Finite Dimensional Case" and published in 1975 by John Wiley and Sons, New York.

I. LINEAR MINIMIZATION ALGORITHMS

In the following pages we shall be concerned mainly with algorithms which estimate the minimum point x_0 of a real valued function $F(x)$ by successively decreasing the value of F along suitably chosen lines.

Let F be a real valued function on a convex set S in an Euclidean space or more generally in a linear space. We assume that F has a minimum point x_0 in S. Starting with an initial point x_1 we seek successive estimates

$$x_{k+1} = x_k + a_k p_k \qquad (k = 1,2,3,\ldots) \tag{1}$$

of x_0, where, at the k-th step, we diminish F along the line

$$x = x_k + \alpha p_k \qquad (-\infty < \alpha < \infty) \tag{2}$$

according to some rule. Here p_k is a direction of descent. Having chosen p_k we normally determine x_{k+1} on the line $x = x_k + \alpha p_k$ by one of the following rules:

(i) Optimal. $\alpha = a_k$ minimizes $\phi_k(\alpha) = F(x_k + \alpha p_k)$.

(ii) Linear Newton. $a_k = \phi_k'(0)/\phi_k''(0)$. We can estimate $\phi_k''(0)$ by the ratio $(\phi_k'(\alpha) - \phi_k(0))/\alpha$, α small.

(iii) Relaxed or buggered. Replace a_k by βa_k in (i) or (ii) where normally $0 < \beta \leq 1$.

(iv) Estimate a_k by a simple rule.

If $\phi_k(\alpha)$ is quadratic in α, Rules (i) and (ii) yield the same value a_k. In the nonquadratic case these rules yield different values of a_k. In the standard gradient methods it is usually helpful to use a relaxation parameter β as described in Rule (iii). As an illustration of Rule (iv) we may select a_k arbitrarily. If $\phi_k(a_k) < \phi_k(0)$ we proceed to the next step. Otherwise we replace a_k by $a_k/2$ and repeat. In the discussion given below we use Rule (ii) unless otherwise specified or implied.

Obviously the success of an algorithm of type (1) is determined in a large part by a judicious choice of the descent vector p_k . Various rules for choosing p_k will be given. The next section will be devoted to the case in which p_k is the negative gradient of F at x_k .

2. GRADIENT METHODS

Perhaps the most obvious choice of the descent vector p_k in the formula $x_{k+1} = x_k + a_k p_k$ is the choice $p_k = -F'(x_k)$, where $F'(x_k)$ is the gradient of F at x_k . Here we have used the symbol $F'(x)$ for the gradient of F at x instead of the more conventional symbol $\nabla F(x)$. Normally the sequence of points $x_1, x_2, x_3, \ldots$ generated by the gradient method converges linearly. However, the convergence may be very slow. Experience has shown that convergence can be accelerated significantly by the introduction of a relaxation factor β as described in Rule (iii), Section 1. For example, the choice $\beta = 0.7$ is usually better than the choice $\beta = 1$. However, I have seen no mathematical theory which specifies how much relaxation, if any, should be used in a given situation.

Gradient methods are not invariant under transformation of variables. For example, under a linear transformation

$$x = x_1 + Uy \tag{3}$$

the gradient $G'(y)$ of

$$G(y) = F(x_1 + Uy)$$

is connected to the gradient F' of F by the relation

$$G'(y) = U^*F'(x_1 + Uy) = U^*F'(x) ,$$

where U^* is the transpose of U. The gradient algorithm

$$y_1 = 0 , \quad y_{k+1} = y_k - a_k G'(y_k)$$

is equivalent, under (3), to the iteration

$$x_{k+1} = x_k - a_k HF'(x_k) , \tag{4}$$

where $H = UU^*$. An algorithm of the form (4) accordingly is also a gradient algorithm. We require that H be positive definite or, at least, that H be nonnegative. When H is a nonnegative matrix of rank $N < n$, Algorithm (4) yields the minimum point of F on the N-plane through x_1 orthogonal to the null space of H. Obviously, we can vary H at each step so as to obtain the algorithm

$$x_{k+1} = x_k - a_k H_k F'(x_k) . \tag{5}$$

This algorithm is often referred to as a *variable metric routine.* This terminology is suggested by the following definition of the gradient as the direction of steepest ascent. Let $F'(x_k,h)$ be the first variation (first differential) of F at x_k. Let h_k be the vector which solves the problem

$$F'(x_k,h) = \max , \quad \|h\|^2 = 1 , \tag{6}$$

where $\|h\|^2$ is a positive definite quadratic form in h. For example, we may select $\|h\|^2 = h^*H_k^{-1}h$. There is a Lagrange multiplier λ_k such that

$$F'(x_k,h) = \lambda_k \langle h_k,h \rangle ,$$

where $2<g,h>$ is the first variation of $\|h\|^2$. Observe that $<g,h>$ is an inner product. Setting $g_k = \lambda_k h_k$ we have

$$F'(x_k,h) = <g_k,h> \tag{7}$$

for all admissible values of h . The choice $<g,h> = g^* H_k^{-1} h$ yields $g_k = H_k F'(x_k)$ as the (generalized) gradient.

Heretofore we tacitly assumed that the points x were points in an Euclidean space. Suppose next that F is a variational integral

$$F(x) = \int_a^b L(t,x(t),\dot{x}(t))dt \ .$$

We seek to minimize F on the class A of arcs $x : x(t)$ $(a \le t \le b)$ joining two fixed points. The first variation $F'(x,h)$ of F along an arc x in A is given by the formula

$$F'(x,h) = \int_a^b \{L_x h + L_{\dot{x}} \dot{h}\}dt \ , \tag{8}$$

where, in the customary variational notation,

$$L_x = \frac{\partial L}{\partial x}(t,x(t),\dot{x}(t)) \ , \quad L_{\dot{x}} = \frac{\partial L}{\partial \dot{x}}(t,x(t),\dot{x}(t)) \ .$$

The variation $h : h(t)$ $(a \le t \le b)$ is constrained to vanish at $t = a$ and $t = b$ so that $h(a) = h(b) = 0$. In order to define the gradient g of F along an arc x in A we introduce the integral

$$\|h\|^2 = \int_a^b |\dot{h}(t)|^2 dt \tag{9}$$

as the square of the norm of h . More generally, we select $\|h\|^2$ to be given by a quadratic integral

$$\|h\|^2 = \int_a^b 2M(t,h(t),\dot{h}(t))dt \ . \tag{10}$$

Then one half of the first variation of $\|h\|^2$ is given by the bilinear form

(11) $$\langle g,h\rangle = \int_a^b \{M_g h + M_{\dot{g}}\dot{h}\}dt .$$

The gradient g of F at x is therefore the variation g such that $g(a) = g(b) = 0$ and

$$F'(x,h) = \langle g,h\rangle$$

for all variations h having $h(a) = h(b) = 0$. In other words, the gradient g solves the differential system

(12) $$(d/dt)M_{\dot{g}} - M_g = (d/dt)L_{\dot{x}} - L_x , \quad g(a) = g(b) = 0 .$$

When $2M = |\dot{h}(t)|^2$ the gradient g is determined by the relations

(13) $$\ddot{g} = (d/dt)L_{\dot{x}} - L_x , \quad g(a) = g(b) = 0 .$$

This formula yields a gradient $g = F'(x)$ of F at x that is suitable for computational purposes. It should be noted that the choice $2M = |h(t)|^2$ does not yield a suitable gradient.

Proceeding in the manner described in the last paragraph we can extend the concept of gradients to more general variational problems and optimal control problems, including multiple integral problems. A further study of such gradients should be fruitful from a theoretical as well as a computational point of view.

3. NEWTON'S METHOD

Newton's method is based on the second order Taylor expansion

$$F(x + h) = Q(x,h) + R(x,h)$$

of F at a point x, where

(14) $$Q(x,h) = F(x) + F'(x,h) + (\tfrac{1}{2})F''(x,h)$$

and $F'(x,h)$, $F''(x,h)$ are the first and second variations (differentials) of F at x. Newton's algorithm proceeds as follows: Starting at a point x_1 we perform the iteration

(15) $$x_{k+1} = x_k + h_k ,$$

where h_k minimizes $Q(x_k,h)$. Thus the problem of minimizing $G(x)$ is reduced to the problem of successively finding the minimum point h_k of the quadratic from $Q(x_k,h)$.

In the n-dimensional case we have

$$Q(x,h) = F(x) + F'(x)^*h + (\tfrac{1}{2})h^*F''(x)h ,$$

where $F'(x)$ is the ordinary gradient of F and $F''(x)$ is the Hessian of F. Clearly

$$h = -F''(x)^{-1}F'(x)$$

minimizes $Q(x,h)$, provided that $F''(x)$ is positive definite, as we shall suppose. Newton's iteration therefore takes the form

(16) $$x_{k+1} = x_k - H(x_k)F'(x_k) , \quad H(x) = F''(x)^{-1}$$

and is accordingly a variable metric routine. If F is of class C'' Newton's method converges superlinearly. The convergence is quadratic when $F''(x)$ is Lipschitzian. In fact, Algorithm (16) converges superlinearly whenever $H(x)$ is continuous at x_0 and $H(x_0) = F''(x_0)^{-1}$, where x_0 is the minimum point of F.

In the case of a variational integral

$$F(x) = \int_a^b L(t,x(t),\dot{x}(t))dt$$

subject to fixed end point constraints, the second variation of F takes the form

$$F''(x,h) = \int_a^b 2M(t,h(t),\dot{h}(t))dt \ , \qquad h(a) = h(b) = 0 \ ,$$

where $2M(t,h,\dot{h})$ is quadratic in h and $\dot{h}$ and is, in fact, the second differential of $L(t,x,\dot{x})$ with respect to x and $\dot{x}$. In this event the minimum point h of $Q(x,h)$ for a fixed arc x is the solution h of the differential system

$$(17) \qquad (d/dt)L_{\dot{x}} - L_x + (d/dt)M_{\dot{h}} - M_h = 0 \ , \qquad h(a) = h(b) = 0 \ ,$$

as one readily verifies. In a similar fashion one can derive Newton's method for general variational problems and optimal control problems.

The main difficulty encountered in Newton's method arises in the determination of the minimum point h_k of the associated quadratic function $Q(x_k,h)$. In the next section we shall discuss effective methods for minimizing a quadratic function.

4. CONJUGATE DIRECTION METHODS

Consider a quadratic function

$$(18) \qquad F(x) = (\tfrac{1}{2})x^*Ax - h^*x + c$$

on an Euclidean space E^n. Here A is a positive definite symmetric matrix and h is a fixed vector. The minimum point x_0 of F is a solution of the linear equation

$$(19) \qquad F'(x) = Ax - h = 0$$

and is accordingly given by the formula $x_0 = A^{-1}h$. In view of the relation

$$F(x) = F(x_0) + (\tfrac{1}{2})(x - x_0)^*A(x - x_0)$$

it is seen that x_0 is the common center of the similar ellipsoids

$$F(x) = b , \quad (b > F(x_0)) .$$

We seek algorithms for finding the minimum point x_0 of F without directly inverting the Hessian A of F. The routines we shall describe are based upon the following fundamental property of a quadratic function F.

The minimum points of F *on parallel lines lie on an* $(n - 1)$*-plane* π_{n-1} *through the minimum point of* F. *The* $(n - 1)$*-plane* π_{n-1} *is said to be conjugate to these lines.*

Restated in terms on ellipsoids we have the following proposition.

The midpoints of parallel chords on an ellipsoid E_{n-1} *lie on an* $(n - 1)$*-plane* π_{n-1} *through the center* x_0 *of* E_{n-1}. *The* $(n - 1)$*-plane* π_{n-1} *is said to be conjugate to these chords.*

On a line $x = x_1 + \alpha p$ through a point x_1 in a direction p, the minimum point $x_2 = x_1 + ap$ of F is the point at which the gradient $F'(x) = Ax - h$ is orthogonal to this line and hence orthogonal to the direction vector p. Consequently $p^*(Ax_2 - h) = 0$ so that x_2 satisfies the equation

$$p^*(Ax - h) = 0 \tag{20}$$

defining an $(n - 1)$-plane π_{n-1} having Ap as its normal. Obviously $x_0 = A^{-1}h$ satisfies this equation so that the minimum point x_0 of F is in π_{n-1}. Since $p^*(Ax_2 - h) = 0$, Equation (20) for π_{n-1} can be put in the form

$$p^*A(x - x_2) = 0 . \tag{21}$$

The $(n - 1)$-plane π_{n-1} is uniquely determined by the direction vector p and does not depend on the initial point x_1. Since Ap is orthogonal to π_{n-1}, the $(n - 1)$-plane π_{n-1} is A-orthogonal to p. *A-orthogonality is called conjugacy*, so that π_{n-1} is conjugate to p. Every vector q in

π_{n-1} is conjugate to p, that is,

$$p^*Aq = 0 . \tag{22}$$

Accordingly we say that two vectors p and q are *conjugate* if $p^*Aq = 0$.

The special property of a quadratic function F described above yields the following conjugate direction algorithm. Starting at an initial point x_1, select a direction p_1 in our n-space π_n and determine the minimum point $x_2 = x_1 + a_1p_1$ of F on the line $x = x_1 + \alpha p_1$ through x_1 in direction p_1. The minimum point x_0 of F on the whole space π_n lies in the $(n - 1)$-plane π_{n-1} through x_2 conjugate to p_1. Accordingly we can restrict our search to the $(n - 1)$-plane π_{n-1} thereby diminishing the dimensionality of our problem by unity. We now repeat the process in the $(n - 1)$-space π_{n-1}. That is, we select a direction p_2 in π_{n-1} and obtain the minimum point $x_3 = x_2 + a_2p_2$ of F on the line $x = x_2 + \alpha p_2$. The minimum point x_0 of F now lies in the $(n - 2)$-plane π_{n-2} in π_{n-1} through x_3 conjugate to p_2, so that again we have diminished the dimensionality of our space of search by unity. Proceeding in this manner, at the k-th step, the point x_0 is known to lie in an $(n - k + 1)$-plane π_{n-k+1} through a previously determined point x_k. Select a direction p_k in π_{n-k+1} and obtain the minimum point $x_{k+1} = x_k + a_kp_k$ of F on the line $x = x_k + \alpha p_k$ through x_k in direction p_k. The $(n - k)$-plane π_{n-k} in π_{n-k+1} through x_{k+1} conjugate to p_k contains the point x_0. Since the dimensionality of our space of search is diminished by one in each step, the point x_{n+1} obtained in the n-th step is the minimum point x_0 of F.

By construction the vector p_k is conjugate to the previously chosen vectors $p_1,\dots,p_{k-1}$, that is,

$$p_j^*Ap_k = 0 , \qquad (j = 1,\dots,k - 1) . \tag{23}$$

Inasmuch as every vector p conjugate to $p_1,\dots,p_{k-1}$ is in π_{n-k}, it follows that the conjugate direction algorithm given above is completely determined by the following two relations:

(24a) x_{k+1} minimizes F on the line $x = x_k + \alpha p_k$;

(24b) p_{k+1} is conjugate to $p_1,\dots,p_k$.

If, for $k = 1,2,3,\dots$, we select p_k to be in the direction of steepest descent of F at x_k on the $(n - k + 1)$-plane π_{n-k+1} through x_k conjugate to $p_1,\dots,p_{k-1}$, we obtain a steepest descent algorithm called the *method of conjugate gradients*. This algorithm is defined by the relations

(25a) x_{k+1} minimizes F on the line $x = x_k + \alpha p_k$;

(25b) $$p_1 = -F'(x_1)\,, \quad p_{k+1} = -F'(x_{k+1}) + \frac{|F'(x_{k+1})|^2}{|F'(x_k)|^2}\, p_k \,.$$

This algorithm generates mutually conjugate vectors $p_1,p_2,p_3,\dots$. Variations of this algorithm will be given in the next section.

An alternative conjugate direction method is the *method of parallel displacements* which can be described as follows. Begin with $n + 1$ independent points $x_1,x_{11},\dots,x_{n1}$. Select $p_1 = x_{11} - x_1$ and minimize F in direction p_1 on parallel lines through $x_{11},\dots,x_{n1}$ to obtain n points $x_2 = x_{12},x_{22},\dots,x_{n2}$. These points determine an $(n - 1)$-plane π_{n-1} which is conjugate to p_1 and which contains the minimum point x_0 of F . We now repeat this algorithm in π_{n-1} . This is done by selecting $p_2 = x_{22} - x_2$ and minimizing F in direction p_2 on parallel lines through $x_{22},\dots,x_{n2}$. We obtain thereby $n - 1$ points $x_3 = x_{23},x_{33},\dots,x_{n3}$ which lie on an $(n - 2)$-plane π_{n-2} containing the minimum point x_0 of F . Next select $p_3 = x_{33} - x_3$ and minimize F in direction p_3 starting from the points $x_{33},\dots,x_{n3}$ and obtaining $n - 2$ new minimum points

$x_4 = x_{34}, x_{44}, \ldots, x_{n4}$ lying on an $(n-3)$-plane π_{n-3} through x_0. Proceeding in this manner we finally obtain the minimum point x_0 of F as the minimum point x_{n+1} of F on the line through x_n and x_{nn}. This procedure is shown schematically in Figure 1 for the case $n = 3$.

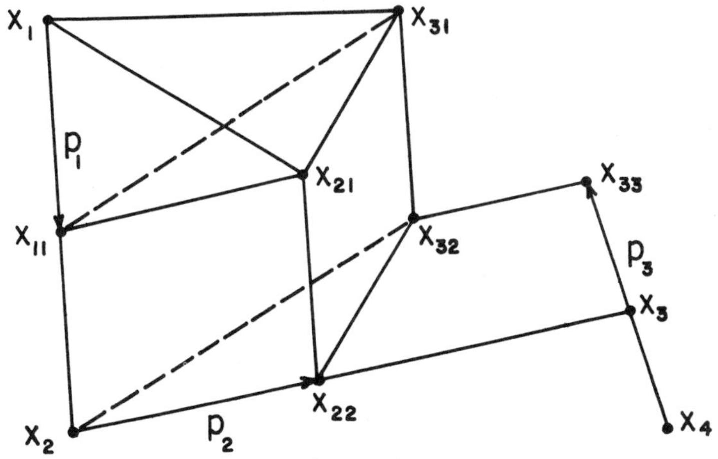

Figure 1.

There is still another geometrical procedure for finding the minimum point x_0 of F which is based on the following proposition.

The minimum points x_k *and* $\bar{x}_k$ *of* F *on distinct parallel* $(k-1)$*-planes* π_{k-1} *and* $\bar{\pi}_{k-1}$ *lie on a line which contains the minimum point* x_{k+1} *of* F *on the* k*-plane* π_k *spanning* π_{k-1} *and* $\bar{\pi}_{k-1}$.

This proposition yields the following algorithm, called the *method of parallel planes*. Starting at an initial point x_1 we obtain the minimum point x_2 of F on a line π_1 through x_1. We next find the minimum point $\bar{x}_2$ of F on a line $\bar{\pi}_1$ parallel to π_1 and obtain the minimum point x_3 of F on the line through x_2 and $\bar{x}_2$. The point x_3 minimizes F on the 2-plane π_2 spanning π_1 and $\bar{\pi}_1$. After obtaining the minimum point $\bar{x}_3$ of F on a 2-plane $\bar{\pi}_2$ parallel to π_2 we find the minimum point x_4 of F on the line joining x_3 to $\bar{x}_3$. Continuing in this manner we find in the n-th step the minimum point $\bar{x}_n$ of F on an $(n - 1)$-plane $\bar{\pi}_{n-1}$ parallel to π_{n-1} and determine the minimum point x_{n+1} of F on the line joining $\bar{x}_n$ to the minimum point x_n of F on π_{n-1}. The point x_{n+1} is the minimum point x_0 of F.

The last algorithm is equivalent to the following routine. Select an initial point x_1 and a vector $p_1 = u_1 \neq 0$. Find the minimum point $x_2 = x_1 + a_1p_1$ of F on the line $x = x_1 + \alpha p_1$. Then perform the iteration defined by the following steps:
(i) Choose u_k so that $u_1,\ldots,u_k$ are linearly independent.
(ii) Set $x_{k1} = x_1 + u_k$ and, for $k = 1,\ldots,k - 1$, obtain the minimum point $x_{k,j+1}$ of F on the line $x = x_{kj} + \alpha p_j$.
(iii) Set $p_k = x_{kk} - x_k$ and determine the minimum point $x_{k+1} = x_k + a_kp_k$ of F on the line $x = x_k + \alpha p_k$. The point x_{n+1} is the minimum point x_0 of F. The $(k - 1)$-planes π_{k-1} and $\bar{\pi}_{k-1}$ described in the preceding algorithm are defined by the equations

$$x = x_1 + \alpha_1p_1 + \cdots + \alpha_{k-1}p_{k-1} ,$$

$$x = x_{k1} + \alpha_1p_1 + \cdots + \alpha_{k-1}p_{k-1} ,$$

respectively. The points x_k and $\bar{x}_k = x_{kk}$ minimize F on π_{k-1} and $\bar{\pi}_{k-1}$ as before. Observe that $p_k = x_{kk} - x_k$ is expressible in the form $p_k = u_k - b_{k1}p_1 - \cdots - b_{k-1}p_{k-1}$ so that $p_1,p_2,\ldots$ are obtained from $u_1,u_2,\ldots$ by a conjugate Gram Schmidt process.

5. CONJUGATE GRADIENT ALGORITHMS

In the last section we gave a geometrical description of the conjugate gradient routine for minimizing a quadratic function

$$F(x) = (\tfrac{1}{2})x^*Ax - h^*x + c ,$$

where the Hessian $A = F''(x)$ of F is positive definite. In addition we gave a formula for generating the conjugate gradients $p_1, p_2, \ldots$. However, we did not give a formula for the scalar a_k appearing in the relation $x_{k+1} = x_k + a_k p_k$. To give precise formulas for carrying out the algorithm, it is convenient to introduce the ngeative gradient of residual

$$r = -F'(x) = h - Ax$$

of F at x . Then the *conjugate gradient algorithm (cg-algorithm)* is defined by the following relations.

(26a) $$x_1 \text{ arbitrary}, \quad r_1 = -F'(x_1) , \quad p_1 = r_1$$

(26b) $$s_k = Ap_k , \quad d_k = p_k^* s_k , \quad c_k = p_k^* r_k$$
$$\text{or} \quad c_k = p_k^* r_1 , \quad a_k = c_k/d_k$$

(26c) $$x_{k+1} = x_k + a_k p_k , \quad r_{k+1} = r_k - a_k s_k$$

(26d) $$p_{k+1} = r_{k+1} + b_k p_k , \quad b_k = -\frac{s_k^* r_{k+1}}{d_k} \quad \text{or} \quad b_k = \frac{|r_{k+1}|^2}{c_k} .$$

Observe that we have given alternative formulas for b_k and c_k . These formulas are unaltered if we replace p_k by $\rho_k p_k$, where ρ_k is a positive scale factor. We have $c_k = |r_k|^2$ as an additional formula for c_k . When a scale factor ρ_k for c_k is used, this formula becomes $c_k = \rho_k |r_k|^2$. In some instances it is useful to select $\rho_{k+1} = 1/(1 + b_k)$ so that the formula for p_{k+1} becomes

$$p_{k+1} = \frac{r_{k+1} + b_k p_k}{1 + b_k} . \tag{27}$$

When this formula is used, the vector p_k is the convex linear combination

$$p_k = \beta_1 r_1 + \cdots + \beta_k r_k , \quad \beta_1 + \cdots + \beta_k = 1$$

of $r_1, \ldots, r_k$ of shortest length. Moreover

$$p_k = -F'(\bar{x}_k) , \quad \bar{x}_k = \beta_1 x_1 + \cdots + \beta_k x_k .$$

It can be shown that $\beta_j = \bar{c}_k / |r_j|^2$ $(j = 1, \ldots, k)$, where $\bar{c}_k = |F'(\bar{x}_k)|^2$. The point $\bar{x}_k$ minimizes the auxiliary quadratic form

$$\hat{F}(x) = (\tfrac{1}{2})|F'(x)|^2$$

on the $(k - 1)$-plane $x = x_1 + \alpha_1 p_1 + \cdots + \alpha_{k-1} p_{k-1}$. The point $\bar{x}_{k+1}$ lies on the line joining $\bar{x}_k$ to x_{k+1} . These facts can be used to obtain useful variations of Algorithm (26) but we shall not pursue these ideas further.

It is of interest to note that for $k > 1$ the conjugate gradient p_k in Algorithm (26) is given by the formula $p_k = x_{kk} - x_k$, where x_{kk} minimizes F on the line $x = x_k + r_k + \alpha p_{k-1}$.

Algorithm (26) can be extended in a number of ways so as to be applicable to nonquadratic functions. We shall give three such extensions.

I. Apply Algorithm (26) with Formula $s_k = Ap_k$ replaced by

$$s_k = \frac{F'(x_1 + \sigma_k p_k) - F'(x_1)}{\sigma_k} , \quad \sigma_k = \frac{\sigma}{|p_k|}$$

where σ is a small positive constant. Restart after n steps with x_{n+1} as the new initial point.

II. Apply Algorithm (26) with formula $s_k = Ap_k$ replaced by

$$s_k = \frac{F'(x_k + \sigma_k p_k) - F'(x_k)}{\sigma_k}, \quad \sigma_k = \frac{\sigma}{|p_k|}$$

where σ is a small positive constant. Select $c_k = p_k^* r_k$ and compute r_{k+1} by the formula $r_{k+1} = -F'(x_{k+1})$. Restart after n steps with x_{n+1} as the new initial point.

III. Apply Algorithm (25) using an appropriate search routine for minimizing F along lines. Restart after n steps with x_{n+1} as the new initial point. This algorithm is known as the Fletcher-Reeves Algorithm.

In each of these algorithms we terminate when $F'(x_{n+1})$ is so small that x_{n+1} can be considered to be a good estimate of the minimum point x_0 of F.

Experience has shown that these algorithms converge at essentially the same rate. Algorithm I requires the fewest number of function evaluations. It is not difficult to show that n steps of Algorithm I with $c_k = p_k^* r_1$ approximates one Newton step. Thus n cg-steps can be viewed to be one Newton step.

In a similar manner the alternative conjugate direction methods given in Section 4 can be modified so as to yield effective routines for minimizing a nonquadratic function.

Finally, it can be shown that the vectors p_k and r_k appearing in Algorithm (26) are connected by the relation

$$p_k = H_{k-1} r_k \tag{28a}$$

where $H_0 = I$ and, for $k = 1,2,\ldots$,

(28b) $$H_k = H_{k-1} - \frac{p_k q_k^* + q_k p_k^*}{d_k} + \left(\frac{\delta_k}{d_k} + 1\right)\frac{p_k p_k^*}{d_k}$$

(28c) $q_k = H_{k-1}s_k$, $s_k = Ap_k$, $\delta_k = q_k^* s_k$, $d_k = p_k^* s_k$.

Here pq^* is the outer product of the vectors p and q. Alternatively we can select

(28b') $$H_k = H_{k-1} - \frac{q_k p_k^*}{\delta_k} + \frac{p_k p_k^*}{d_k}$$

in which case $p_{k+1} = H_k r_{k+1}$ is equivalent to Formula (27). Obviously Algorithm (26) and the extended Algorithms I, II and III can be modified by using Formulas (28) to compute the vectors $p_1, p_2, \ldots$. When this is done in Algorithms II and III it is no longer necessary to restart after n steps. The resulting algorithms are called variable metric algorithms and are also referred to as Davidon-Fletcher-Powell routines. The term "variable metric" is used because of the relation

$$x_{k+1} = x_k - a_k H_{k-1} F'(x_k) .$$

Algorithm (26) is applicable without change to a positive definite quadratic function F on a Hilbert space having x^*y as its inner product. In this event the points $x_1, x_2, x_3, \ldots$ generated by the algorithm converge superlinearly to the minimum point x_0 of F. A cg-algorithm can be applied whenever gradients and inner products are well-defined as in the case of the fixed end point variational problem described above. A detailed study of these and similar situations should be a fruitful avenue of research.

6. AUGMENTABILITY AND THE METHOD OF MULTIPLIERS

Constrained minimum problems play a significant role in applications of optimization theory. It is important to devise effective methods for obtaining numerical solutions of problems of this type. One such method is the method of multipliers which is based on the concept of augmentability which we shall now explain.

Consider the problem of minimizing a function $f(x)$ subject to a single constraint

$$g(x) = 0 . \tag{29}$$

The extension to a set of constraints

$$g_j(x) = 0 \qquad (j = 1,\ldots,m)$$

is obvious. One needs only to consider g to be an m vector and make appropriate interpretations in our formulas. We suppose that x_0 is a solution of our problem. Clearly x_0 also minimizes any augmented function F of the form

$$F = f + \lambda g + (\sigma/2)g^2 \tag{30}$$

subject to the constraint $g(x) = 0$. We shall say that f is *augmentable* at x_0 if the multiplier λ and the constant σ can be chosen so that x_0 affords an unrestricted minimum to F . If we set

$$L = f + \lambda g , \qquad G = (\tfrac{1}{2})g^2$$

so that

$$F = L + \sigma G$$

and if F has an unrestricted minimum at x_0 , then, using the "del" notation for the gradient with respect to x , we have

$$\nabla F(x_0) = \nabla L(x_0) = 0 , \qquad F''(x_0,h) = L''(x_0,h) + \sigma G''(x_0,h) \geq 0$$

for every vector $h \neq 0$. Inasmuch as

$$G''(x_0,h) = g'(x_0,h)^2$$

we have the first and second order Lagrange multiplier rule

$$\nabla L(x_0) = 0 \ , \quad L''(x_0,h) \geq 0$$

$$\text{whenever} \quad h \neq 0 \quad \text{and} \quad g'(x_0,h) = 0 \ .$$

Conversely, if σ can be chosen so that the Lagrange multiplier rule holds with " $\geq$ " replaced by " $>$," there is a positive number σ such that

$$F''(x_0,h) = L''(x_0,h) + \sigma G''(x_0,h) > 0 \quad \text{whenever} \quad h \neq 0$$

so that F has at least a local minimum at x_0 . Consequently, the strengthened Lagrange multiplier rule implies (local) augmentability of f at x_0 . The multiplier λ eliminates the first order terms in F at x_0 and the constant σ convexifies F .

The concept of augmentability suggests the following algorithm for finding the minimum point x_0 of f subject to $g = 0$. In our discussion we use the notation $F(x,\lambda,\sigma)$ to signify the dependence of F on x , λ and σ . The problem at hand is to find the multiplier λ and a constant σ so that x_0 affords an unrestricted minimum to F . Obviously if a constant σ_0 is effective for x_0 so also is any $\sigma > \sigma_0$. The *method of multipliers* we propose proceeds as follows. Choose an initial multiplier λ_1 , a reasonably large constant σ and a positive number ξ_0 . For $k = 1,2,3,\ldots$ select $\xi_k \geq \xi_0$, obtain an approximate minimum point x_k of $F(x,\lambda_k,\sigma + \xi_k)$, and update λ by the formula

$$\lambda_{k+1} = \lambda_k + \xi_k g(x_k) \ . \tag{31}$$

If convergence is slow increase σ and/or ξ_0 . Alternative updating formulas have been given by various authors.

The method of multipliers is a stable numerical method. An alternative method, which may be unstable numerically, is the method of penalty functions in which x_k minimizes $F(x,0,\sigma_k)$, where $\lim \sigma_k = \infty$. Under favorable conditions the sequence $\{x_k\}$ converges to x_0 and $\{\sigma_k g(x_k)\}$ converges to the associated multiplier λ. The method of penalty functions is an excellent tool for deriving the first order Lagrange multiplier rule without assumptions of augmentability or of normality. A discussion of these ideas can be found in my recent book, "Optimization Theory, The Finite Dimensional Case."

Augmentability and the associated method of multipliers is not restricted to finite dimensional problems. Consider, for example, the Problem of Lagrange in which we minimize an integral

$$I(x) = \int_a^b L(t,x(t),\dot{x}(t))dt$$

on the class A of arcs $x : x(t)$, $(a \le t \le b)$ joining two fixed points and satisfying the differential constraint

$$\phi(t,x(t),\dot{x}(t)) = 0 .$$

If a particular arc $x_0 : x_0(t)$, $(a \le t \le b)$ satisfies the standard sufficiency conditions for minimizing I on A, there exists a multiplier $\lambda(t)$ and a function $\sigma(t,x,\dot{x})$ such that x_0 affords a local minimum to the augmented integral

$$J(x,\lambda,\sigma) = \int_a^b \{L + \lambda\phi + \frac{\sigma}{2}\phi^2\}dt$$

on A with the constraint $\phi = 0$ removed. Accordingly we say that I is *augmentable at* x_0 if there is an augmented integral $J(x,\lambda,\sigma)$ of this type such that x_0 affords a minimum to J on A with the constraint $\phi = 0$ removed. Again augmentability leads to a method of multipliers in which $\lambda(t)$ is updated by the formula

$$\lambda_{k+1}(t) = \lambda_k(t) + \xi_k \phi(t, x_k(t), \dot{x}_k(t)) .$$

In a similar manner the simple control problem in which

$$I = \int_0^T L(t, x(t), u(t))dt$$

is minimized on the class of functions $x(t)$, $u(t)$, $(0 \leq t \leq T)$ satisfying the constraints

$$\dot{x} = f(t,x,u) , \qquad x(0) , x(T) \text{ prescribed}$$

can be augmented by an integral of the form

$$J = \int_0^T \{L + p(\dot{x} - f) + \frac{\sigma}{2}(\dot{x} - f)^2\}dt$$

when the standard sufficiency conditions hold, thereby eliminating the constraint $\dot{x} = f$. These and other more general variational and optimal control problems have been studied by Russell Rupp, who has established convergence theorems for the associated methods of multipliers.

Differential Equations

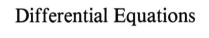

A CRITIQUE OF OSCILLATION OF DIFFERENTIAL SYSTEMS

W. T. Reid

Austin, Texas

1. INTRODUCTION

In the theory of differential equations the term "oscillation" has been used in a multiplicity of manners. The purpose of the present discussion is to survey the array of definitions of this concept, and to consider some of their respective functional aspects. In only a minor fashion will there be concern with specific criteria for oscillation and/or non-oscillation. Moreover, attention will be limited to the case of finite dimensional ordinary differential equations and systems with real-valued coefficients, and wherein the domain of the independent variable is an interval on the real line. For treatments of various aspects of the theory, and including many problems excluded from the present consideration, the reader is

ISBN 0-12-632250-3

referred to such works as Hartman [12], Swanson [43], Hille [16], Coppel [7], Reid [36], Kreith [21] and Friedland [9].

Historically, the topic is founded on the study of the nature of real solutions of a real linear homogeneous second order ordinary differential equation appearing in the Sturmian theory for such equations. Specifically, we shall be concerned largely with the generalizations of this concept to vector Hamiltonian systems, emanating from the work of Marston Morse [23,24], and for higher order equations results that have grown from the basic results of Leighton and Nehari [22] for fourth order equations.

2. SECOND ORDER EQUATIONS

The linear homogeneous second order equation to be considered will be written as

$$\mathcal{l}[u](t) \equiv [r(t)u'(t) + q(t)u(t)]' - [q(t)u'(t) + p(t)u(t)] = 0 , \tag{2.1}$$

where it is assumed that the coefficient functions satisfy the following hypothesis.

(H) On a given interval I on the real line, r, p and q are real-valued with $r(t) > 0$ and $r, 1/r, p, q$ belonging to $L^{\infty}_{loc}(I)$.

In particular, these conditions are implied by the more classical assumption:

(H_C) On a given interval I the real-valued functions r,p,q are continuous, with r positive.

In either case, the concept of a solution of (2.1) is that of a solution $(u;v)$ in the Carathéodory sense of the first order system $ru' + pu = v$, $v' = qu' + pv$, which may be written as

$$(2.2)\qquad \begin{aligned} \ell_1[u,v](t) &\equiv -v'(t) + c(t)u(t) - a(t)v(t) = 0 , \\ \ell_2[u,v](t) &\equiv u'(t) - a(t)u(t) - b(t)v(t) = 0 , \end{aligned}$$

wherein $a = -q/r$, $b = 1/r$, $c = p - q^2/r$. In general, whenever reference is made to a solution u of (2.1) the associated function v is understood to be $ru' + qu$. The class of equations (2.1) or (2.2) satisfying hypothesis (H) on I is denoted by $\Lambda[I]$.

Of particular interest are equations (2.1) with $q(t) \equiv 0$. Indeed, whenever hypothesis (H) is satisfied and $w(t) = \exp\{-\int_a^t [q(s)/r(s)]ds\}$ with $a \in I$, then under the transformation $u = wy$ equation (2.1) becomes

$$(2.1')\qquad [r_1(t)y'(t)]' - p_1(t)y(t) = 0 ,$$

wherein $r_1 = rw^2$ and $p_1 = w^2[p - q^2/r]$; equivalently, under $u = wy$, $v = z/w$ system (2.2) is reduced to

$$(2.2')\qquad \begin{aligned} -z'(t) + c_1(t)y(t) &= 0 , \\ y'(t) - b_1(t)z(t) &= 0 , \end{aligned}$$

with $b_1 = 1/r_1 = b/w^2$, $c_1 = p_1 = w^2c$.

If $s \in I$, a value $s_1 \neq s$ on I is said to be *conjugate* to s , [with respect to (2.1) or (2.2)], if there exists a non-identically vanishing solution u of (2.1) satisfying $u(s) = 0 = u(s_1)$. The concept of conjugate point apparently goes back to Weierstrass, who in his 1879 lectures introduced the term in a corresponding situation for the Jacobi equation in a plane parametric variational problem. Correspondingly, if $s \in I$ and (u,v) is a solution of (2.2) satisfying the initial conditions $u(s) \neq 0$, $v(s) = 0$, then a value $s_1 \in I$ is called a *right-hand*, (*left-hand*) *focal point* to $t = s$ if $s_1 > s$, ($s_1 < s$), and $u(s_1) = 0$. The English term *focal* appeared in the early 1900's and dates from A. Kneser's use of *Brennpunkte* in the consideration of the Jacobi equation for a parametric variational problem with an end-point restricted to lie on a given surface, and wherein the variational "transversality condition" was equivalent to orthogonality. In view of the reduction of (2.2) to (2.2') as described above, it follows readily that if s_1 and s_2, ($s_1 < s_2$), are point of I which are conjugate with respect to (2.2) then there exists a value $s_3 \in (s_1, s_2)$ such that s_2 is a right-hand focal point to s_3 and s_1 is a left-hand focal point to s_3.

Historically, it is of interest to note that in 1909 Picone [29; Sec. 1] introduced further terms for a more extended classification of points relative to solutions of a second order equation

$$u''(t) + p(t)u'(t) + q(t)u(t) = 0,$$

with real continuous coefficients p,q on an interval I. For a given $s \in I$, let $u_0(t)$ and $u_1(t)$ be solutions of this equation satisfying the initial conditions $u_0(s) = 0$, $u_0'(s) \neq 0$, $u_1(s) \neq 0$, $u_1'(s) = 0$. In Picone's terminology, a value $t = s_1$ is called a *right-hand*, (*left-hand*) *pseudo-conjugate* {pseudoconuigato} of $t = s$ if $s_1 > s$, ($s_1 < s$) and $u_0'(s_1) = 0$. Also, a value $t = s_1$ is called a *right-hand* (*left-hand*) *hemiconjugate* {emiconuigato} to $t = s$ if $s_1 > s$, ($s_1 < s$), and $u_1(s_1) = 0$. Finally, Picone called $t = s_1$ a *right-hand* (*left-hand*) *deconjugate* {deconuigato} to $t = s$ if $s_1 > s$, ($s_1 < s$), and $u_1'(s_1) = 0$. Thus in Picone's terminology a focal point to $t = s$ is a hemiconjugate to s. As far as the author is aware, following Picone's introduction of this terminology there was no use of this classification. Indeed, it appears that the only subsequent comprehensive terminology corresponding to that of Picone is due to Borůvka, [6; §3 of Ch. I], who uses the term "conjugate point of the first, second, third and fourth class," with the associated right- or left-hand designations, for the respective Picone conjugate, deconjugate, pseudoconjugate, and hemiconjugate.

Returning to the consideration of conjugate points, for $s \in I$ let $u(t) = u(t;s)$, $v(t) = v(t;s)$ be the solution of (2.2) satisfying $u(s) = 0$, $v(s) = 1$. Then the zeros of $u(t)$ are isolated, and for $\varphi_o(s) = s$ and $\varphi_k(s)$, $(\varphi_{-k}(s))$, the k-th zero of $u(t)$ following, (preceding), the value s we have the monotone increasing sequence

$$(2.4) \qquad \{\varphi_j(s)\}, \quad (j = 0,\pm1,\pm2,\ldots)$$

of consecutive conjugate points on I. Either of the subsequences $\{\varphi_k(s)\}$ or $\{\varphi_{-k}(s)\}$, $(k = 1,2,\ldots)$, may be vacuous, finite, or infinite. In the following statements of properties of the sequence (2.4) we shall assume that both of these subsequences are infinite, since whenever one of them is finite or vacuous the corresponding alteration in statement is obvious.

A. Algebraic Properties

1. $\varphi_{-j}[\varphi_j(s)] = \varphi_o(s) = s$, $(j = 0,\pm1,\pm2,\ldots)$;
2. $\varphi_{i+j}(s) = \varphi_i[\varphi_j(s)]$, $(i,j = 0,\pm1,\pm2,\ldots)$; in particular, $\varphi_{1+j}(s) = \varphi_1[\varphi_j(s)]$.

That is, $\{\varphi_j(s)\}$ is an Abelian group with group operation $\circledast$ given by $\varphi_i \circledast \varphi_j = \varphi_i[\varphi_j]$.

B. Analytic Properties

For $h = \pm1,\pm2,\ldots$ the domain of definition of $\varphi_h(s)$ is a subinterval $I_h(s)$ of I on which

1_h. φ_h is strictly monotone increasing;

2_h. φ_h is continuous.

F. Functional Properties

For $[a,b]$ a compact subinterval of I, let $D[a,b]$ denote the class of real-valued absolutely continuous functions y with $y' \in L^2[a,b]$, and $D_0[a,b] = \{y : y \in D[a,b], y(a) = 0 = y(b)\}$. Then

$$J[y|a,b] = \int_a^b (ry'^2 + 2qyy' + py^2)dt \tag{2.4}$$

is a quadratic functional on $D[a,b]$. Moreover, if there exists a real-valued solution $(u_0;v_0)$ of (2.2) with $u_0(t) \neq 0$ on $[a,b]$, then

$$J[y|a,b] = (v_0/u_0)y^2\Big|_a^b + \int_a^b r[y' - (u_0'/u_0)y]^2 dt,$$

$$\text{for } y \in D[a,b];$$

in particular, when such a solution $(u_0;v_0)$ exists $J[y|a,b]$ is positive definite on $D_0[a,b]$.

1. If $b = \varphi_1(a)$ then $J[y|a,b] \geq 0$ for $y \in D_0[a,b]$, and $J[y|a,b] = 0$ iff $y(t) = cu(t;a)$ for some constant c. Moreover, one has the separation result that if $u(t)$ is an arbitrary real-valued non-identically vanishing solution of (2.2) then either $u(t)$ is of the form $cu(t;a)$ or there exists a value $s_0 \in (a,b)$ such that $u(s_0) = 0$.
2. For $k = 1,2,\ldots$ we have $\varphi_k(a) = b$ iff one of the following conditions holds:

   (i) On $D_0[a,b]$ the quadratic functional $J[y|a,b]$

has negative index $n^-[a,b]$ equal to $k - 1$, and non-negative index $n^o[a,b]$ equal to k.

(ii) $\lambda = 0$ is the k-th eigenvalue of the two-point boundary problem

$$\mathbb{B}[a,b] \qquad \mathcal{L}[u](t) + \lambda u(t) = 0 , \quad u(a) = 0 = u(b) .$$

3. If $r^\alpha, p^\alpha, q^\alpha$, $(\alpha = 1,2)$, are functions satisfying (H) on I, and for arbitrary non-degenerate compact subintervals $[a,b]$ of I we have that $J^{1,2}[y|a,b] = J^2[y|a,b] - J^1[y|a,b]$ is non-negative {positive} definite on $D_o[a,b]$, then for $s \in I$ the corresponding sequences $\{\varphi_j^\alpha(s)\}$, $(\alpha = 1,2)$, satisfy the following inequalities:

$$(2.5) \qquad \begin{aligned} &\varphi_k^1(s) \le \varphi_k^2(s) , \quad \{\varphi_k^1(s) < \varphi_k^2(s)\} , \quad (k = 1,2,\dots) , \\ &\varphi_{-k}^1(s) \ge \varphi_{-k}^2(s) , \quad \{\varphi_{-k}^1(s) > \varphi_{-k}^2(s)\} . \end{aligned}$$

D. Disconjugacy

Equation (2.1), or system (2.2), is said to be *disconjugate* on a subinterval I_o of I if there exists no pair of conjugate points on I_o.

1. (2.1) is disconjugate on I_o iff for $[a,b] \subset I_o$ the functional $J[y|a,b]$ is positive definite on $D_o[a,b]$.
2. If $s \in I$ and $\varphi_1(s)$ exists, then $\varphi_1(s)$ is the supremum of values $c \in I^+(s) = \{t : t \in I, t \ge s\}$ such that (2.1) is disconjugate on $[s,c]$; in case $\varphi_1(s)$ is an interior point of I, then also $\varphi_1(s)$ is the infimum, (actually minimum), of values $b \in I^+(s)$ such that (2.1) is not disconjugate on $[a,b]$.

3. For $I_0 \subset I$ the class $\mathbb{D}(I_0)$ of equations (2.1) with coefficients satisfying (H), and disconjugate on I_0, possesses the following properties:

(i) $\mathbb{D}(I_0)$ is a positive convex cone; i.e., with multiplication by scalars and addition defined in the obvious manners, if $\ell \in \mathbb{D}(I_0)$ and $c > 0$ then $c\ell \in \mathbb{D}(I_0)$, while if $\ell^\alpha \in \mathbb{D}(I_0)$, $(\alpha = 1,2)$, then $\ell^1 + \ell^2 \in \mathbb{D}(I_0)$. In particular, $\theta\ell^1 + (1 - \theta)\ell^2 \in \mathbb{D}(I_0)$ for $0 \leq \theta \leq 1$.

(ii) If for $[a,b] \subset I$ and $\ell^j \in \Lambda[I]$, $(j = 1,2)$, $d(\ell^1,\ell^2)$ is defined as the greatest of the $L^\infty[a,b]$ norms of $r^1 - r^2$, $(1/r^1) - (1/r^2)$, $p_1 - p_2$ and $q_1 - q_2$, then in the metric space $\{\Lambda[a,b]:d(\cdot,\cdot)\}$ the set $\mathbb{D}[a,b]$ is open and connected.

O. Oscillation on Non-Compact Intervals

If I is a non-compact interval of the form $[a_0,\omega)$, $-\infty < a_0 < \omega \leq \infty$, a non-identically vanishing solution of (2.1) is called *oscillatory on* I if on this interval it has infinitely many zeros, and *non-oscillatory* in the contrary case. An equation (2.1) is called *non-oscillatory* if all of its non-identically vanishing solutions are non-oscillatory, and *oscillatory* if at least one of its solutions is oscillatory. Also, (2.1) is termed *eventually disconjugate* on $I = [a_0,\omega)$ in the case there exists a $c \in I$ such that the equation is disconjugate on $[c,\omega)$. In view of the separation result of F.1 it follows that for an equation (2.1) either all non-identically vanishing solutions are oscillatory on $[a_0,\omega)$ or all such solutions are non-oscillatory on this interval. In particular, we have the following properties of equations (2.1):

1. Eventual disconjugacy implies non-oscillation.
2. Non-oscillation implies eventual disconjugacy.

3. OSCILLATION THEORY OF SELF-ADJOINT HAMILTONIAN SYSTEMS

Morse [23,24] generalized the concept of conjugate point, and extended the basic oscillation and comparison theorems of the Sturmian theory to vector equations

$$\begin{aligned} L[u](t) \equiv {} & [R(t)u'(t) + Q(t)u(t)]' \\ & - [Q^*(t)u'(t) + P(t)u(t)] = 0 , \end{aligned} \tag{3.1}$$

wherein, similar to the conditions in Section 2, it is assumed that on a given interval I on the real line the $n \times n$ matrix functions P,Q,R satisfy the following hypothesis.

(H) P,Q,R are real-valued, $R(t)$ and $P(t)$ are symmetric with $R(t) > 0$, (i.e., positive definite) for $t \in I$, and R,R^{-1},P,Q are of class $L^{\infty}_{loc}(I)$.

Corresponding to (2.1) and (2.2), $u(t)$ is termed a solution of (3.1) if $u(t),v(t) = R(t)u'(t) + P(t)u(t)$ is a solution in the Carathéodory sense of the first order "Hamiltonian system"

$$\begin{aligned} L_1[u,v](t) &\equiv -v'(t) + C(t)u(t) - A^*(t)v(t) = 0 , \\ L_2[u,v](t) &\equiv u'(t) - A(t)u(t) - B(t)v(t) = 0 , \end{aligned} \tag{3.2}$$

where A,B,C are the $n \times n$ matrix functions $A = -R^{-1}Q$, $B = R^{-1}$, $C = P - Q^*R^{-1}Q$. In particular, in view of (H) we have

$(\hat{H}\text{-}1)$ $\quad B(t) \equiv B^*(t)\ ;\quad C(t) \equiv C^*(t)\ ;\quad B(t) \geq 0$

$$\text{for}\quad t \in I\ ;\quad A,B,C \in L^{\infty}_{loc}(I)\ .$$

Now the non-singularity of $B(t)$ on I implies the following property of systems (3.2) equivalent to (3.1) in the above sense.

$(\hat{H}\text{-}2)$ System (3.2) is identically normal on I ; i.e., if $(u(t) \equiv 0, v(t))$ is a solution on a non-degenerate subinterval I_o of I then also $v(t) \equiv 0$ on I_o and $(u(t);v(t)) \equiv (0,0)$ on I .

Hypothesis $(\hat{H}\text{-}2)$ clearly does not require $B(t)$ to be non-singular so that the study of systems (3.2) satisfying hypotheses $(\hat{H}\text{-}1,2)$ is potentially more general than the consideration of equations (3.1). Such is indeed the case. In particular, if $p_0(t),\ldots,p_n(t)$ are real-valued functions with $p_n(t) > 0$ and $p_\alpha(t) \in C^\alpha(I)$, then the 2n-th order self-adjoint scalar differential equation

$$(3.3)\qquad \mathcal{L}_{2n}[u](t) \equiv \sum_{\alpha=0}^{n} (-1)^\alpha [p_\alpha(t) u^{[\alpha]}(t)]^{[\alpha]}$$

is equivalent under the transformation

$$u_j(t) = u^{[j-1]}(t)\ ,\ (j = 1,\ldots,n)\ ;\quad v_n(t) = p_n(t) u^{[n]}(t)\ ;$$
$$v_{n-k}(t) = p_{n-k}(t) u^{[n-k]}(t) - v'_{n-k+1}(t)\ ,\ (k = 1,\ldots,n-1)\ ,$$

to a system (3.2), wherein $A_{\alpha\beta}(t) \equiv 0$ unless $\beta = \alpha + 1$, $A_{\alpha,\alpha+1}(t) \equiv 1$, while $B(t)$ and $C(t)$ are diagonal matrices with $B_{\alpha\alpha}(t) \equiv 0$ for $\alpha = 1,\ldots,n-1$, $B_{nn}(t) = 1/p_n(t)$, $C_{\beta\beta}(t) = p_{\beta-1}(t)$, $(\beta = 1,\ldots,n)$, and hypotheses $(\hat{H}\text{-}1,2)$ hold.

If hypothesis $(\hat{H}\text{-}1)$ is satisfied, and $(u_1;v_1),(u_2;v_2)$ are two solutions of (3.2), then the function $v_2^*(t)u_1(t) - u_2^*(t)v_1(t)$ is constant on I. Following a terminology introduced by von Escherich in 1898 for such systems appearing as the canonical form of the accessory equations for a variational problem of so-called Lagrange or Bolza type, if the value of this constant is zero the solutions $(u_1;v_1),(u_2;v_2)$ are said to be *(mutually) conjugate*. If $(U;V)$ is a $2n \times h$ matrix function whose column vectors are linearly independent solutions of (3.2) with $V^*U - U^*V \equiv 0$, these solutions form a basis for a conjugate family of solutions of dimension h, consisting of all linear combinations of these column vectors. The maximum dimension of a conjugate family of solutions of (3.2) is n, and a given family of dimension $h < n$ is contained in a conjugate family of dimension n. If U and V are $n \times n$ matrix functions such that the column vectors of $(U;V)$ form a basis for an n-dimensional conjugate family of solutions of (3.2), then for brevity $(U;V)$ is called a *conjugate base* for this system. In particular, if $s \in I$ then each of the solutions $(U(t;s);V(t;s))$ and $(U_0(t;s);V_0(t;s))$ of the matrix system $L_1[U,V] = 0$, $L_2[U,V] = 0$ determined by the initial conditions

$$\begin{aligned} U(s;s) &= 0\ ,\ V(s;s) = E\ ; \\ U_0(s;s) &= E\ ,\ V_0(s;s) = 0\ ; \end{aligned} \tag{3.4}$$

is a conjugate base for (3.2).

Corresponding to the definition of conjugate point for a second order scalar equation in Section 2, distinct values s and s_1 of I are said to be *(mutually) conjugate*, with respect to (3.2), if there exists a solution $(u;v)$ of this system with $u(s) = 0 = u(s_1)$ and $u(t) \not\equiv 0$ on the interval with endpoints s and s_1. If both $(\hat{H}\text{-}1)$ and $(\hat{H}\text{-}2)$ are satisfied, then for $s \in I$ a distinct value $s_1 \in I$ is conjugate to s iff the conjugate base $(U(t;s);V(t;s))$ defined by (3.4) is such that $U(s_1;s)$ is singular. Moreover, if $U(s_1;s)$ is of rank $n - r$ then there exist r linearly independent vectors $\xi^{(j)}$, $(j = 1,\ldots,r)$, such that $u_j(t) = U(t;s)\xi^{(j)}$, $v_j(t) = V(t;s)\xi^{(j)}$ are linearly independent solutions of (3.2) determining s_1 conjugate to s; in this case r is called the *index* of s_1 as a conjugate point to s. *Throughout the following discussion we shall consider only systems* (3.2) *satisfying both* ($\hat{H}$-1) *and* ($\hat{H}$-2). In case the system is not identically normal, the determination of points conjugate to s is no longer through the consideration of a single conjugate base, (see, for example, [36; Secs. 3, 4 of Ch. VII, and Problem VII.4.3] and [38]). Indeed, for the treatment of oscillation phenomena in the abnormal case there need be a modification of the notion of conjugate point, in essence replacing it by a "conjugate interval" concept, (see, for example, [33; Sec. 4].

In displaying the sequence of values conjugate to a given $s \in I$ it becomes reasonable to repeat each value a number of times equal to its index, so that corresponding to (2.4) we now have a sequence $\{\varphi_j(s)\}$, $(j = 0,\pm1,\pm2,\ldots)$, where

$$(3.5) \quad \varphi_0(s) = s \, , \; \varphi_j(s) \le \varphi_{j+1}(s) \, , \; (j = 0,\pm1,\pm2,\ldots) \, .$$

Again, we shall proceed as though both of the subsequences $\{\varphi_k(s)\}$, $\{\varphi_{-k}(s)\}$, $(k = 1,2,\ldots)$, are infinite, although either may be finite or vacuous. With the sequence of conjugate points thus defined we have satisfied the algebraic property A.1 $\varphi_{-j}[\varphi_j(s)] = \varphi_0(s) = s$, although the proof is far from trivial as in the case of (2.2). However, Property A.2 is not valid in general; in its place we have

A.2^0 $\varphi_{j+k}(s) \leq \varphi_k[\varphi_j(s)]$, $(j = 0,\pm1,\pm2,\ldots;k = 1,2,\ldots)$. Also, for each $h = \pm1,\pm2,\ldots$ for which $\varphi_h(s)$ has a non-vacuous domain of definition its domain is an interval and Conditions B.1_h and B.2_h are satisfied.

For $[a,b]$ a compact subinterval of I , let $D[a,b]$ denote the class of real-valued vector functions y absolutely continuous on $[a,b]$ and $y \in D[a,b]:z$ in the sense that there exists an associated $z \in L^2[a,b]$ such that $L_2[y,z] = 0$ on $[a,b]$; moreover $D_0[a,b] = \{y : y \in D[a,b] , y(a) = 0 = y(b)\}$. Corresponding to the functional properties of Section 2, we now have that

$$J[y|a,b] = \int_a^b (z^*Bz + y^*Cy)dt \tag{3.6}$$

is a quadratic functional on $D[a,b]$, and if $(U_0;V_0)$ is a conjugate base for (3.2) with $U_0(t)$ non-singular on $[a,b]$, then

$$J[y|a,b] = y^*V_0U_0^{-1}y \Big|_a^b + \int_a^b (z - VU^{-1}y)^*B(z - VU^{-1}y)dt$$

$$\text{for } y \in D[a,b]:z .$$

In particular, when such a conjugate base $(U_0;V_0)$ exists, $J[y|a,b]$ is positive definite on $D_0[a,b]$.

F.1 If $b = \varphi_1(a)$, then $J[y|a,b] \geq 0$ for $y \in D_0[a,b]$, and $J[y|a,b] = 0$ iff there exists a solution $(u;v)$ of (3.2) determining b as $\varphi_1(a)$ with $u(t) = y(t)$ on $[a,b]$. Moreover, if $(U;V)$ is any conjugate base for (3.2) then either there exists a non-zero n-dimensional vector ξ such that $u(t) = U(t)\xi$, $v(t) = V(t)\xi$ is a solution of (3.2) determining $\varphi_1(a)$ conjugate to a , (i.e., $u(a) = 0 = u(b)$), or there exists a value $s_0 \in (a,b)$ such that $U(s_0)$ is singular. (Reid [38; Thm. 3.1].)

F.2 For $k = 1,2,\ldots$ we have $\varphi_k(a) = b$ iff one of the following conditions holds:

(i) There exists a $j \in \{1,\ldots,k\}$ such that on $D_0[a,b]$ the quadratic functional $J[y|a,b]$ has negative index $n^-[a,b]$ equal to $j-1$, and non-negative index $n^0[a,b]$ not less than k.

(ii) $\lambda = 0$ is the k-th eigenvalue of the two-point boundary problem

$$L_1[u,v](t) = \lambda u(t)\ , \quad L_2[u,v](t) = 0\ ,$$

$$\mathbb{B}[a,b] \qquad u(a) = 0 = u(b)\ .$$

The fact that the span of the eigenfunctions of $\mathbb{B}[a,b]$ corresponding to negative (non-positive) eigenvalues is a maximal subspace on which $J[y|a,b]$ is negative (non-positive) definite is a special instance of the concept of natural isoperimetric conditions introduced by Birkhoff and Hestenes [4]. Now for $c \in I$, $c > a$, let $D_0^I[a,c]$ denote the class of functions $y \in D(I)$ with $y(a) = 0$, $y(t) \equiv 0$ for $t \geq c$, so that if $b \in I$, $b > a$, then for $c \in (a,b)$ the restriction of an element of $D_0^I[a,c]$ to $[a,b]$ is an element of $D_0[a,b]$. In particular, if $c > a$ is a conjugate point to a and $(u_c;v_c)$ is a solution of (3.2) determining c as conjugate to a, let y_c be the element of $D_0^I[a,c]$ satisfying $y_c(t) = u_c(t)$ on $[a,c]$. Then the span of the y_c for $c \in (a,b)$, ($c \in (a,b]$), is also a maximal subspace of $D_0[a,b]$ on which $J[y|a,b]$ is negative, (non-positive), definite, and therefore an individual $c_0 \in (a,b)$ is conjugate to a iff c_0 is a point of discontinuity of the negative index $n^-[a,c]$ of $J[y|a,b]$ on $D_0^I[a,c]$, and the index of c_0 as a conjugate point to a is equal to the jump of $n^-[a,c]$ at c_0. These facts provided a major stimulus for the abstract theory of oscillation emanating from the paper [15] of Hestenes on the application of the theory of quadratic forms in Hilbert space to the problem of oscillation.

Now if $(U;V)$ is any conjugate base for (3.2) a value c at which $U(c)$ is singular, and of rank $n - r$, is called a *focal point* of $(U;V)$ of index r. The points of singularity of $U(t)$ are isolated, and if $a \in I$ is a point of non-singularity of $U(t)$, then for $b > a$, $b \in I$, the number of focal points of $(U;V)$ on (a,b), ($(a,b]$), with each counted a number of times equal to its index, is equal to the negative (non-positive) index of the quadratic functional

$$J_a[y|a,b] = y^*(a)[V(a)U^{-1}(a)]y(a) + J[y|a,b]$$

on the class of y belonging to $D_{*o}[a,b] = \{y : y \in D[a,b] , y(b) = 0\}$. Results corresponding to those stated above for conjugate points also hold for focal points in general, and the interrelations between such results provide extensions of the oscillation and comparison theorems of classical Sturmian theory to such results for the conjugate bases for a system. In particular, one of the fundamental results of this theory is the following, (see, for example, Morse [23; Cor. 1 to Thm. 6] or Reid [36; Cor. 1 to Thm. VII.7.9]).

(iii) For a given subinterval I_o of I the number of focal points on I_o of any conjugate base for (3.2) differs from that of any other conjugate base by at most n.

F.3 Suppose that for $\alpha = 1,2$ the $n \times n$ matrix functions $A^\alpha(t),B^\alpha(t),C^\alpha(t)$ satisfy hypotheses $(\hat{H}\text{-}1,2)$; moreover, if $D^\alpha[a,b]$, $(\alpha = 1,2)$, denote the respective classes $D[a,b]$ for the corresponding systems (3.2), then $D^1[a,b] = D^2[a,b]$ for arbitrary $[a,b] \subset I$ and $J^{1,2}[y|a,b] = J^2[y|a,b] - J^1[y|a,b]$ is non-negative (positive) definite on $D^1_o[a,b] = D^2_o[a,b]$. Then for $s \in I$ we have for the corresponding sequences $\{\varphi_j^{\alpha}(s)\}$, $(\alpha = 1,2)$, the inequalities (2.5). In view of F.2-ii above, this result is a ready consequence of Theorem 12.1 in Ch. VII of [36]; for related results, and discussion of conditions under which $D^1[a,b] = D^2[a,b]$ for arbitrary $[a,b] \subset I$, see also [38; Sec. 4].

As for scalar equations (2.1) or (2.2), a system (3.2) is said to be disconjugate on a subinterval I_o of I if there exists no pair of conjugate points on I_o. In view of the above comments and Properties F.1 and F.2, it follows readily that statements D.1 and D.2 of Section 2 remain valid for systems satisfying ($\hat{H}$.1,2).

In particular, if (3.2) is the system derived from an equation (3.1) with coefficient matrix functions satisfying (H), and $v(t) = R(t)u'(t) + Q(t)u(t)$, then for $[a,b] \subset I$ the class $D[a,b]$ is the class of n-dimensional vector functions $y(t)$ absolutely continuous on $[a,b]$ and with $y' \in L^2[a,b]$. Moreover, $J[y|a,b]$ of (3.6) is then also equal to

$$\int_a^b (y^{*\prime}[Ry' + Qy] + y^*[Q^*y' + Py])dt .$$

Consequently, if for I_o a subinterval of I the class of systems (3.2) disconjugate on I_o is denoted by $\mathbb{D}(I_o)$, on the subclass of such systems derived from equations (3.1) satisfying (H) the properties corresponding to D.3-i,ii may be established by the same method as indicated in Section 2, (see, in particular, Coppel [7; Ch. 2, Sec. 5]). Indeed, similar results are derivable in like manner for self-adjoint equations (3.3) with $p_\alpha \in C^\alpha(I)$ and $p_n(t) > 0$ on I, since for the related system (3.2) an n-dimensional vector function y belongs to $D[a,b]$ if and only if $y_\beta(t) = \xi^{[\beta-1]}(t)$, $(\beta = 1,\dots,n)$, with $\xi(t) \in C^{[n-1]}[a,b]$ and $\xi^{[n-1]}$ absolutely continuous having $\xi^{[n]} \in L^2[a,b]$, while the corresponding functional $J[y|a,b]$ is equal to $\int_a^b \Sigma_{\alpha=0}^n p_\alpha(t)[\xi^{[\alpha]}(t)]^2 dt$.

For still more general systems (3.2) satisfying ($\hat{H}$-1,2) one may proceed as in [35; Sec. 3] to consider for each integer q, $0 < q \leq n$, the subset I_q of I on which $B(t)$ has rank $n - q$, and choose on I_q an $n \times q$ measurable matrix $\pi(t,q)$ satisfying $B(t)\pi(t,q) = 0$, $\pi^*(t,q)\pi(t,q) = E_q$. Then the $(n + q) \times (n + q)$ matrix

$$\begin{bmatrix} B(t) & \pi(t,q) \\ \pi^*(t,q) & 0 \end{bmatrix}, \quad t \in I_q, \tag{3.7}$$

is non-singular, and its inverse is of the form

$$\begin{bmatrix} \Gamma(t,q) & \pi(t,q) \\ \pi^*(t,q) & 0 \end{bmatrix}, \quad t \in I_q. \tag{3.8}$$

The $n \times n$ matrix $\Gamma(t,q)$ is of rank $n - q$ on I_q, and $\Gamma(t,q) \geq 0$. The $n \times n$ matrix function defined on $[a,b]$ as equal to $\Gamma(t,q)$ on I_q is the E. H. Moore generalized reciprocal $B^{\#}(t)$ of $B(t)$, and $B^{\#}(t) \geq 0$ for $t \in [a,b]$. Moreover, if $\pi(t)$ is defined as equal to the $n \times q$ matrix $\pi(t,q)$ on I_q, then whenever $y \in D[a,b]:z$ we have

$$\pi^*(t)L_o[y](t) = 0, \quad t \in [a,b],$$

and

$$z^*Bz + y^*Cy = (L_o[y])^*\Gamma L_o[y] + y^*Cy,$$

$$\text{where } L_o[y] = y' - Ay.$$

Conditions ($\hat{H}$-1,2) are not strong enough to imply that $\Gamma(t,q)$ is integrable on compact subsets of I_q, and in general the matrix function $B^{\#}(t)$ is not integrable on $[a,b]$. However, $B^{\#} \in L^{\infty}[a,b]$ whenever the non-negative symmetric matrix function $B(t)$ satisfies the following condition (see Reid [35; p. 281]).

($\hat{H}$-3) For $[a,b] \subset I$ there exists a positive constant $k_o = k_o[a,b]$ such that $B^2(t) - k_o B(t) \geq 0$ for t a.e. on $[a,b]$.

In particular, this condition is satisfied by $B(t)$ belonging to systems (3.2) derived from an equation (3.1) for which (H) holds, or systems (3.2) representing self-adjoint equations (3.3) with coefficient function satisfying the conditions specified above.

Now consider a maximal class S of systems (3.2) satisfying ($\hat{H}$-1,2,3), and such that if $A^\alpha, B^\alpha, C^\alpha$, $(\alpha = 1,2)$, are two sets of coefficient matrix functions belonging to systems of S then $B^1(t)$ and $B^2(t)$ have equal ranks almost everywhere on I, with a common $n \times q$ matrix $\pi(t,q)$ satisfying $B^1(t)\pi(t,q) = B^2(t)\pi(t,q) = 0$ on I_q, and $\pi^*(t,q)A^1(t) = \pi^*(t,q)A^2(t)$ almost everywhere on I_q; algebraically, these conditions may be expressed as $\mathrm{rk}\,[B^1(t) \;\; B^2(t)] = \mathrm{rk}\, B^1(t) = \mathrm{rk}\,[B^1(t) \;\; A^2(t) - A^1(t)]$ almost everywhere on I. If $I_o \subset I$, then the systems of S which are disconjugate on I_o possess properties which are direct extensions of D.3-i,ii of Section 2. Indeed, if $A^\alpha, B^\alpha, C^\alpha$ belong to two such systems, and $\Gamma^\alpha(t,q)$, $(\alpha = 1,2)$, are the corresponding matrix functions of (3.8) for $B(t) = B^\alpha(t)$ in (3.7), then for $\theta \in [0,1]$ the matrix function $\Gamma_\theta(t,q) = (1 - \theta)\Gamma^1(t,q) + \theta\Gamma^2(t,q)$ is of rank $n - q$ and satisfies $\Gamma_\theta(t,q) \geq 0$ on I_q. Consequently, if $\Gamma_\theta(t) = \Gamma_\theta(t,q)$ for $t \in I_q$ and $B_\theta(t) = [\Gamma_\theta(t)]^\#$, then the system (3.2) with coefficient matrix functions $(1 - \theta)A^1 + \theta A^2$, B_θ, $(1 - \theta)C^1 + \theta C^2$ satisfies hypotheses ($\hat{H}$-1,2,3). Moreover, since the classes $D^\alpha_o[a,b]$ for the respective problems are the same, and the corresponding functionals J^1, J^2, J_θ of the

$\mathbb{B}[\varphi_j^o(s),\varphi_{j+1}^o(s)]$ of Section 3. In view of the continuity and monotonity of the eigenvalues $\lambda_\alpha = \lambda_\alpha[a,b]$ of $\mathbb{B}[a,b]$ as functions of the interval $[a,b]$ one has the following results.

F.2° For $k = 1,2,\ldots$, if $b > \varphi_k^o(a)$, $b \in I$, and $N(k) = \Sigma_{\beta=0}^{k-1} n_\beta(a)$ then $\lambda_{N(k)}[a,b] < 0$; also, if $n_k'(a)$ denotes the minimum of the values $n_o(a),\ldots n_{k-1}(a)$, and $N_1(k) = (k - 1)n + n_k'(a) + 1$, then $\lambda_{N_1(k)}[a,\varphi_k^o(a)] \geq 0$. For the case of second order scalar equations considered in Section 2 we have $n = 1$, $n_\beta(a) = 1$ for $\beta = 0,1,\ldots,k - 1$, and the above relations reduce to the following statements: $\lambda_k[a,b] < 0$ if $b > \varphi_k(a)$, $b \in I$, and $\lambda_{k+1}[a,\varphi_k^o(a)] > 0$.

5. HIGHER ORDER DIFFERENTIAL EQUATIONS

Within recent years considerable attention has been given to oscillation phenomena for higher order linear homogeneous differential equations

$$(5.1) \qquad \mathcal{L}_m[u](t) \equiv p_m(t)u^{[m]}(t) + \cdots + p_o(t)u(t) = 0 , \quad t \in I ,$$

with coefficient functions $p_j(t)$, $(j = 0,1,\ldots,m)$, real-valued, continuous, and $p_m(t) \neq 0$ on an interval I on the real line. In the majority of cases treatments have emanated from the basic paper of Leighton and Nehari [22] on fourth order equations of the form

$$(5.2) \qquad [r(t)u''(t)]'' - p(t)u(t) = 0 , \quad t \in [a,\infty) ,$$

where r and p are real-valued continuous functions with $r(t) > 0$ and $p(t)$ of constant sign on $I = [a,\infty)$.

In the early 1960's basic work in this area was published by the Russian mathematicians Aliev, A. Ju Levin, Azbelev and

Caljuk. Subsequently, a great many writers of papers in this area have lineage with either the mathematical group headed by Nehari at Carnegie-Mellon, (notably, M. Hanan, B. Schwartz, U. Elias), or those headed by John Barrett at the University of Utah and the University of Tennessee, (notably, W. J. Coles, R. W. Hunt, T. L. Sherman, A. J. Zettl, D. B. Hinton, J. M. Dolan, A. C. Peterson, J. W. Heidel, G. B. Gustafson, J. R. Ridenhour, G. W. Johnson). In the following discussion, with a very few exceptions, for references to work and bibliography of papers previous to 1969, the reader is referred to the posthumous monograph of Barrett [3], or to Chapter 3 and bibliography of Coppel's *Disconjugacy* [7].

In a terminology closely related to that of [22], but with some evolutionary deviation, an equation (5.1) is called *disconjugate* on a subinterval I_o of I in case any non-identically vanishing real solution of this equation has on I_o at most $m - 1$ zeros, where each zero is counted according to multiplicity. If $s \in I$ and for a positive integer k there exist values $b \in I^+(s) = \{t : t \in I, t \geq s\}$ such that there are real-valued non-identically vanishing solutions of (5.1) with at least $m - 1 + k$ zeros on $[s,b]$, each zero counted according to multiplicity, then the infimum of such values b is called the k-th *right-hand conjugate point* to s, with respect to (5.1), and denoted by $\eta_k(s)$. An *extremal solution* of (5.1) defining $\eta_k(s)$ is a non-identically vanishing real-valued solution of this equation with zeros at $t = s$ and $t = \eta_k(s)$, and having at least $m - 1 + k$ zeros on $[s,\eta_k(s)]$. In a similar manner one may define the k-th left-hand conjugate point $\eta_{-k}(s)$ of s. If we set $\eta_o(s) = s$ then, as in the earlier sections, for $s \in I$ we have a uniquely determined sequence of conjugate points

$$\{\eta_j(s)\}\ , \quad (j = 0,\pm 1,\pm 2,\dots)\ ,$$

where, as before, either of the individual sequences $\{\eta_k(s)\}$, $\{\eta_{-k}(s)\}$, $(k = 1,2,\dots)$, may be infinite, finite, or vacuous. In particular, (5.1) is disconjugate on a subinterval I_o of I iff for $s \in I_o$ we have $\eta_{\pm 1}(s) \notin I_o$.

Although for $m = 2$ the above definitions of conjugate points are equivalent to the earlier ones of Section 2, such is not the case in general for $m > 2$. In particular, if (5.1) is a self-adjoint equation of the form (3.3) with $m = 2n$ and $n > 1$, the concept of conjugate point is quite distinct from that defined in Section 4 in terms of an associated first order system (3.2). As far as techniques are concerned, in the context of Section 4 the study of disconjugacy centers around extremizing properties of the associated quadratic functional $J[y|a,b]$. For the work discussed in this section, however, many of the basic results on disconjugacy involve utilization of the mean value results of Polya [30], and the fact that an equation (5.1) is disconjugate on an open interval $I_o \subset I$ iff there exist positive functions $r_\alpha(t)$, $(\alpha = 1,\dots,m+1)$, of respective classes $C^{m+1-\alpha}(I_o)$ such that $\mathcal{L}_m[u]$ has a (Frobenius-Polya-Mammana) factorization

$$\mathcal{L}_m[u] = r_{m+1}\frac{d}{dt}r_m\frac{d}{dt}\cdots\frac{d}{dt}r_1 u\ , \tag{5.3}$$

and the associated study of the oscillation properties of specific "principal solutions" of (5.1). In this connection, the reader is referred to Nehari [25], Coppel [7; Ch. 3], and Hartman [13].

In the early 1960's, A. Ju Levin and T. L. Sherman showed that if on I there exists a pair of conjugate points with respect to (5.1) then the domain of existence of $\eta_1(s)$ is a subinterval of I on which this conjugate point function is strictly monotone increasing and continuous; also, there exists a non-identically vanishing real-valued solution $u(t)$ of (5.1) which has a zero of order at least r, $(1 \leq r \leq m-1)$, at $t = s$ and a zero of order at least $m - r$ at $t = \eta_1(s)$, while $u(t) \neq 0$ for $t \in (s, \eta_1(s))$.

The study of higher order conjugate point functions $\eta_k(s)$, $k \geq 2$, and the characterization of the distribution of zeros of corresponding extremal solutions, falls into two classes: consideration limited to certain types of fourth order equations, and the treatment of such matters for general m-th order equations. In the first category belong results of Leighton and Nehari [22], Peterson [28] and Pudei [31,32] on the continuity of general $\eta_k(s)$ for certain fourth order equations, and the results of Ridenhour and Sherman [39] on the characterization of the zeros of extremal functions for fourth order equations under the assumption that certain types of distributions of zeros are excluded for all solutions of the equation. In the second category, the papers of Ridenhour [41] and Keener [19,20] are concerned with conditions in terms of boundary-value functions under which one may obtain a characterization of extremal functions for $\eta_k(s)$ analogous to a characterization given by Leighton and Nehari [22] for a fourth order equation. For equations of the particular form

$$\mathcal{L}_m[u](t) + p(t)u(t) = 0 , \tag{5.4}$$

wherein $\mathcal{L}_m[u]$ is a linear differential form such that the corresponding equation (5.1) is disconjugate on $I = [a, \infty)$, and $p(t)$ is a continuous function which does not change sign on I, the most general result appears to be that obtained recently by Elias [8]. He has shown that if $u(t)$ is an extremal solution of (5.4) for $[s, \eta_k(s)]$ then $u(t)$ has exactly

$m - 1 + k$ zeros on this interval, and the only zeros of $u(t)$ on the open interval are $k - 1$ zeros of odd multiplicity; also, the zero of $u(t)$ at $\eta_k(s)$ is of odd or even multiplicity according as $p(t) \geq 0$ or $p(t) \leq 0$ on I. Moreover, if the domain of existence of η_k is non-vacuous then this domain is of the form $[0,b)$, $0 < b \leq \infty$, and η_k is a strictly increasing continuous function on this interval. These results are extensions of earlier work of Johnson [17,18] who considered the case of m even.

Theorem 1 of Ridenhour [40] provides a necessary and sufficient condition for the continuity of $\eta_k(s)$, $(k > 1)$, for general equations (5.1); application of this theorem usually requires knowledge of the nature of the extremal solutions for $\eta_k(s)$. In the general area of considerations of the continuity of higher order conjugate point functions, of particular interest is the recent manuscript of Gustafson and Ridenhour [11], in presenting an example of a fourth order linear equation of the form $(p(t)u''')' + q(t)u'' = 0$ for which the conjugate point functions $\eta_2(s)$ and $\eta_3(s)$ are discontinuous.

As far as results for equations (5.1) corresponding to those of D.3-i,ii of Section 2, it is to be noted that the totality of equations (5.1) which are disconjugate on a given open interval $I_0 \subset I$ form a connected open set in a suitable normed function space. This follows from the fact that if $\mathcal{L}_m^\beta[u] = 0$, $(\beta = 1,2)$, are two such equations, and in the factorizations (5.3) the respective values of the $r_\alpha(t)$, $(\alpha = 1,\ldots,m + 1)$, are denoted by $r_\alpha^\beta(t)$, then for $\theta \in [0,1]$ and $r_\alpha^\theta(t) = (1 - \theta)r_\alpha^1(t) + \theta r_\alpha^2(t)$ the equation $\mathcal{L}^\theta[u] = 0$ having factorization (5.3) with $r_\alpha(t) = r_\alpha^\theta(t)$ is also disconjugate on I_0.

A result in the same general area, but different in character is given in Theorem 4.1 of Nehari [25], to the effect that if $q_1(t)$ and $q_2(t)$ are continuous functions satisfying $0 \le q_1(t) \le q_2(t)$ on I, and both (5.1) and the equation $\mathcal{X}_m[u](t) + q_2(t)u(t) = 0$ are disconjugate on an open subinterval I_o of I, then the equation $\mathcal{X}_m[u](t) + q_1(t)u(t) = 0$ is also disconjugate on I_o.

In regard to the question of oscillation on a non-compact interval $[a,\omega)$, Gustafson [10] has given an example of a self-adjoint fourth order equation of the form $[r(t)u''(t)]'' + [q(t)u'(t)]' = 0$ with r,q continuous and $r(t) > 0$ on an interval $[a,\infty)$ which is non-oscillatory, but which is not eventually disconjugate. Indeed, on an arbitrary $[c,\infty) \subset [a,\infty)$, and arbitrary positive integer N, there is a solution of this equation with N successive double zeros, so that even in the more restrictive variational sense the equation is not eventually disconjugate. Finally, it is to be noted that for general equations (5.1) Ridenhour [42] has presented conditions which imply that non-oscillation is equivalent to eventual disconjugacy; the conditions are in the form of assumptions that certain boundary-value functions are infinite for all values of the argument.

6. OTHER DEFINITIONS OF OSCILLATION FOR DIFFERENTIAL SYSTEMS

Consider a vector differential equation of the form

$$(6.1) \qquad [R(t)u'(t) + Q_1(t)u(t)]' - [Q_2^*(t)u'(t) + P(t)u(t)] = 0 ,$$

with real-valued $n \times n$ matrix coefficient functions, wherein $R(t)$ is non-singular and R, R^{-1}, Q_1, Q_2, P are of class $L^\infty_{loc}(I)$, but without assumptions of symmetry of R, P and

$Q_1 = Q_2$ as in (3.1). For such a system one may then define conjugate point in precisely the same manner as before; that is, distinct values s and s_1 are conjugate if there exists a non-identically vanishing solution $u(t)$ with $u(s) = 0 = u(s_1)$. As the equation is no longer self-adjoint, there is no longer available the important concept of a conjugate base. However, one may still derive certain disconjugacy conditions with the aid of comparison results involving (6.1) and the associated self-adjoint equation

(6.2) $$[R^{o}(t)u'(t) + Q_3(t)u(t)]' - [Q_3^{*}(t)u'(t) + P^{o}(t)u(t)] = 0 ,$$

where $Q_3 = \frac{1}{2}[Q_1 + Q_2]$, $R^{o} = \frac{1}{2}[R + R^{*}]$, $P^{o} = \frac{1}{2}[P + P^{*}]$. For an equation (6.1) with $Q_1 = Q_2 = 0$, results of this sort were obtained by Hartman and Wintner [14]. Also, for an equation essentially of the form (6.1), and with complex-valued coefficients, corresponding results were presented in Reid [34; Sec. 5]. For matrix equations analogous to (6.1) with $n = 4$, a more detailed study has been given by Barrett [2].

For a system of the form

(6.3) $$\begin{aligned} -v'(t) + C(t)u(t) - D(t)v(t) &= 0 , \\ u'(t) - A(t)u(t) - B(t)v(t) &= 0 , \end{aligned}$$

wherein the matrix functions A,B,C,D are now of respective dimensions $n \times n$, $n \times m$, $m \times n$, $m \times m$ and belong to $L_{loc}(I)$, Reid [37; Ch. II] has maintained a definition of conjugate point similar to that for the self-adjoint system (3.2). That is, values s,s_1 of I, $s < s_1$, are called conjugate with respect to (6.3) if there exists a solution $(u;v)$ of this system with $u(s) = u(s_1) = 0$ and $u(t) \not\equiv 0$ on $[s,s_1]$. In particular, this definition is such that if $n < m$, and s,s_1 are arbitrary distinct values on I, then there exists a non-identically vanishing solution $(u;v)$ of (6.3) satisfying

$u(s) = 0 = u(s_1)$. However, this fact does not imply that s and s_1 are conjugate since all such solutions $(u;v)$ of (6.3) may have $u(t) \equiv 0$.

A quite different type of oscillation has been introduced by Nehari [26] for first order systems

$$y'(t) = A(t)y(t) , \tag{6.4}$$

where $A(t)$ is an $n \times n$ real-valued matrix function on an interval I . A non-identically vanishing solution $y(t)$ of (6.3) is said to be oscillatory on I if each of its components assumes the value zero at some point of I ; i.e., if there exist values $t_\alpha \in I$, $(\alpha = 1,\dots,n)$, such that $y_\alpha(t_\alpha) = 0$, $(\alpha = 1,\dots,n)$. The equation is said to be oscillatory on I if it possesses at least one oscillatory solution on this interval, and non-oscillatory in the contrary case. For such a system the (first) right-hand conjugate point $\eta_1(s)$ of a value s is defined as the smallest value $b > s$ for which there exists a non-identically vanishing solution y of (6.4) such that each of its components vanish at either $t = s$ or $t = b$. For systems (6.4) that can be reduced to a linear equation of the form (5.1), the first conjugate point $\eta_1(s)$ as defined in Section 5 does indeed have this property. For general systems (6.4), however, it is shown by an example that for $\eta_1(s)$ defined as above it is not necessary for $\eta_1(s)$ to be the infimum of values c such that $[s,c)$ is an interval of oscillation of the system. If no point of I possesses a conjugate point, (6.4) is said to be disconjugate on I . Whereas an equation which is non-oscillatory on I is also disconjugate on this interval, the converse is in general not true. In a subsequent paper [27], Nehari derived conditions, expressed in terms of various norms of A , which guarantee the non-oscillation of (6.4) on a given interval.

For first order systems of the form (2.2) or (3.2) it is to be noted that the above definitions of intervals of oscillation and conjugate points blend the concepts of conjugate points and focal points as considered in Sections 2 and 3. For example, for the second order equation $u'' + u = 0$, and equivalent first order system $u' = v$, $v' = -u$, in the terminology of Section 2 the first right-hand conjugate point to $t = 0$ is $t = \pi$, with determining solution $u = \sin t$, $v = \cos t$. On the other hand, according to the definition of [26] for the first order system, the first right-hand conjugate point to $t = 0$ is $t = \pi/2$, with two independent determining solutions $u = \sin t$, $v = \cos t$ and $u = \cos t$, $v = -\sin t$.

7. CONCLUDING REMARKS

From the above survey it is clear that there remains a number of unsolved problems for future consideration. Especially for higher order self-adjoint equations, there are questions concerning more intimate relationships between the concepts of oscillation as presented in Section 3 and in Section 6. In the area of conjugacy in the sense of Section 6, the recent results of Gustafson and Ridenhour make even more intriguing the question of continuity of higher order conjugate point functions. For problems in this domain there remains a basic question as to the possible existence of functional techniques of fundamental importance. In particular, do there exist functionals which might assume a role equal in importance to that occupied by quadratic functionals in the context of Section 3?

From the standpoint of terminology, it appears unfortunate that the word "conjugate" has been so overworked. The concept as introduced in Section 3 might be segregated by the adjunction of the adjective "variational," but this modification is not truly descriptive. In all cases the matter is one of the solvability of allied boundary problems, and indeed in the consideration of boundary problems involving two-point boundary conditions at varying end-points a and b, some authors have called these values conjugate, with respect to the considered problem, whenever the boundary problem has a non-trivial solution. In particular, such terminology has been used by Atkinson [1; Ch. X] for self-adjoint two-point boundary problems involving a differential system equivalent to (3.2). I have no firm suggestion as to the solution of this problem of terminology, although one possibility would be to introduce a system of discriminating symbols similar to the manner in which $i_1 - i_2 - \cdots - i_k$ and $r_{i_1 i_2 \cdots i_k}$ have become accepted in the literature dealing with the oscillation theory of Section 6. However, I do feel that the proliferation of usages of the term "conjugate" is a verbal pollution of the mathematical literature to which I have unfortunately contributed personally, and it is to be hoped that we will strive to be mathematical environmentalists in the search to decrease, if not remove, this pollution.

BIBLIOGRAPHY

1. Atkinson, F. V. "Discrete and Continuous Boundary Problems." Academic Press, New York, 1964. MR 31#416.

2. Barrett, J. H. Systems-disconjugacy of a fourth order differential equation, *Proc. Amer. Math. Soc.* *12*(1961), pp. 205-213. MR 24#A304.

3. Barrett, J. H. Oscillation theory of ordinary linear differential equations, *Advances in Mathematics* *3*(1969), pp. 415-509. MR 41#2113.

4. Birkhoff, G. D. and M. R. Hestenes. Natural isoperimetric conditions in the calculus of variations, *Duke Math. J. 1* (1935), pp. 198-286.

5. Bliss, G. A. and I. J. Schoenberg. On separation, comparison and oscillation theorems for self-adjoint systems of linear second order differential equations, *Amer. J. Math. 53*(1931), pp. 781-800.

6. Borůvka, O. "Lineare Differentialtransformationen zweiter Ordnung." Veb. Deutsche Verlag der Wissenschaften, Berlin (1967). MR 38#4743; English translation, English Univ. Press, London, 1971. MR 52#549.

7. Coppel, W. A. "Disconjugacy." Lecture Notes in Mathematics No. 220, Springer-Verlag, 1971.

8. Elias, U. The extremal solutions of the equation $Ly + p(x)y = 0$, I; II, *J. Math. Analy. Appl. 50*(1975), pp. 447-457; *55*(1976), pp. 253-265.

9. Friedland, S. Nonoscillation, disconjugacy and integral inequalities, Amer. Math. Soc. Memoirs, No. 176, 1976.

10. Gustafson, G. B. Eventual disconjugacy of self-adjoint fourth order linear differential equations, *Proc. Amer. Math. Soc. 35*(1972), pp. 187-192. MR 45#7178.

11. Gustafson, G. B. and J. R. Ridenhour. Lower order branching and conjugate function discontinuities, mms.

12. Hartman, P. "Ordinary Differential Equations." John Wiley and Sons, New York, 1960. MR 30#1270.

13. Hartman, P. Principal solutions of disconjugate n-th order linear differential equations, *Amer. J. Math. 91*(1969), pp. 306-362. MR 40#450.

14. Hartman, P. and A. Wintner. On disconjugate differential equations, *Can. J. Math. 8*(1956), pp. 72-81. MR 17-611.

15. Hestenes, M. R. Applications of the theory of quadratic forms in Hilbert space in the calculus of variations, *Pacific J. Math. 1*(1951), pp. 525-581. MR 13-759.

16. Hille, E. "Lectures on Ordinary Differential Equations." Addison Wesley, Reading, Mass., 1969. MR 40#2939.

17. Johnson, G. W. The k-th conjugate point function for an even order linear differential equation, *Proc. Amer. Math. Soc.* *42*(1974), pp. 563-568. MR 48#11665.

18. Johnson, G. W. Conjugate point properties for an even order linear differential equation, *Proc. Amer. Math. Soc.* *45*(1974), pp. 371-376. MR 50#684.

19. Keener, M. S. Oscillatory solutions and multi-point boundary value functions for certain n-th order linear ordinary differential equations, *Pacific J. Math.* *51*(1974), pp. 187-202. MR 50#5096.

20. Keener, M. S. On the equivalence of oscillation and the existence of infinitely many conjugate points, *Rocky Mtn. J. Math.* *5*(1975), pp. 125-134. MR 50#10437.

21. Kreith, K. "Oscillation Theory." Lecture Notes in Mathematics No. 324, Springer-Verlag, 1973.

22. Leighton, W. and Z. Nehari. On the oscillation of solutions of self-adjoint linear differential equations of the fourth order, *Trans. Amer. Math. Soc.* *89*(1958), pp. 325-377. MR 21#1429.

23. Morse, M. A generalization of the Sturm separation and comparison theorems in n-space, *Math. Ann.* *103*(1930), pp. 52-69.

24. Morse, M. "The Calculus of Variations in the Large." Amer. Math. Colloq. Publ., Vol. 18, Amer. Math. Soc., New York, 1934.

25. Nehari, Z. Disconjugate linear differential operators, *Trans. Amer. Math. Soc.* *129*(1967), pp. 500-516. MR 36#2860.

26. Nehari, Z. Oscillation theorems for systems of linear differential equations, *Trans. Amer. Math. Soc.* *139*(1969), pp. 339-347. MR 39#542.

27. Nehari, Z. Nonoscillation and disconjugacy of systems of linear differential equations, *J. Math. Analy. Appl.* *42* (1973), pp. 237-254. MR 47#8976.

28. Peterson, A. C. The distribution of zeros of extremal solutions of a fourth order differential equation for the n-th conjugate point, *J. Differential Eqs.* *8*(1970), pp. 502-511. MR 42#4821.

29. Picone, M. Su un problema al contorno nelle equazioni differenziali lineari ordinarie del secondo ordine, *Ann. Scuola Norm. Sup. Pisa 10*(1909), pp. 1-92.

30. Polya, G. On the mean-value theorem corresponding to a given linear homogeneous differential equation, *Trans. Amer. Math. Soc.* *24*(1924), pp. 312-324.

31. Pudei, V. Properties of solutions of the differential equation $y^{(4)} + p(x)y'' + q(x)y = 0$, [Czech], *Časopis Pěst. Mat.* *93*(1968), pp. 201-216. MR 38#3515.

32. Pudei, V. Über die Eigenschaften der Lösungen linearer Differentialgleichungen gerader Ordnung, *Časopis Pěst. Mat.* *94*(1969), pp. 401-425. MR 42#7993.

33. Reid, W. T. Boundary value problems of the calculus of variations, *Bull. Amer. Math. Soc.* *42*(1937), pp. 633-666.

34. Reid, W. T. Oscillation criteria for linear differential systems with complex coefficients, *Pacific J. Math.* *6*(1956), pp. 733-751. MR 18-898.

35. Reid, W. T. Variational methods and boundary problems for ordinary linear differential equations, Proc. U.S.-Japan Sem. on Differential and Functional Equations, Univ. of Minnesota, Minneapolis, Minn., June 26-30, 1967. W. A. Benjamin, Inc., pp. 267-299. MR 37#4332.

36. Reid, W. T. "Ordinary Differential Equations." John Wiley and Sons, New York, 1971. MR 42#7963.

37. Reid, W. T. "Riccati Differential Equations." Academic Press, New York, 1972. MR 50#10401.

38. Reid, W. T. A supplement to oscillation and comparison theory for Hermitian differential systems, *J. Differential Eqs.* *16*(1974), pp. 550-573.

39. Ridenhour, J. R. and T. L. Sherman. Conjugate points for fourth order linear differential equations, *SIAM J. Appl. Math.* *22*(1972), pp. 599-603. MR 46#2151.

40. Ridenhour, J. R. On the continuity of conjugate point functions, *SIAM J. Appl. Math.* *27*(1974), pp. 531-538. MR 51#3609.

41. Ridenhour, J. R. On the zeros of solutions of N-th order linear differential equations, *J. Differential Eqs.* *16* (1974), pp. 45-71. MR 51#1012.

42. Ridenhour, J. R. Linear differential equations where non-oscillation is equivalent to eventual disconjugacy, *Proc. Amer. Math. Soc.* *49*(1975), pp. 366-372. MR 51#1013.

43. Swanson, C. A. "Comparison and Oscillation Theory of Linear Differential Equations." Academic Press, New York, 1968.

CONJUGATE POINTS WITHOUT THE CONDITION OF LEGENDRE

Everett Pitcher

Lehigh University
Bethlehem, Pennsylvania

1. THE DIFFERENTIAL EQUATION AND CONJUGATE FAMILIES

The differential equation is the general self-adjoint system in m variables, namely,

$$L(\eta) \equiv \frac{d}{dx}(R\eta' + Q\eta) - (Q^T\eta' + P\eta) = 0 \ . \tag{1.1}$$

Here R , Q , P are $m \times m$ matrices of real valued functions of one variable, at least continuous and smooth enough for the operations subsequently to be performed. The aspect of the problem to be investigated appears even with quite smooth coefficients. The matrices R and P are assumed symmetric. Symbols ^{T} and $'$ denote transpose and derivative, respectively, and η is a column of m rows, the unknown in the equation. With

$$2\omega(\eta,\zeta) = \zeta^T R\zeta + \zeta^T Q\eta + \eta^T Q^T\zeta + \eta^T P\eta \tag{1.2}$$

one finds that

ISBN 0-12-632250-3

$$L(\eta) = \frac{d}{dx}\omega_\zeta(\eta,\eta') - \omega_\eta(\eta,\eta') \tag{1.3}$$

so that (1.1) is the Euler equation of integrals with $2\omega(\eta,\eta')$ as integrand. A solution of (1.1) is sometimes called an *extremal*.

The matrix $R(x)$ is assumed to be nonsingular. In the classical problem, by virtue of the hypotheses of positive regularity, the quadratic form $z^T R(x) z$ is positive definite. Positive regularity is not assumed here.

By the *negative index*, usually called *index*, (resp., positive index) of a quadratic function $q : X \to \mathbb{R}$ is meant the least upper bound $h_q[1_q]$, possibly $+\infty$, of the dimension of linear spaces $Y \subset X$ such that $q|Y$ is negative (positive) definite. With $q(z) = b(z,z)$ where $b : X \times X \to \mathbb{R}$ is bilinear, the characteristic space N of q is the linear space $N = \{z \mid b(z,w) = 0 \ \forall \ w \in X\}$. The dimension of N is the *nullity* of q.

The problem under consideration is said to be of *type* ρ if $z^T R(x) z$ is a quadratic function of index ρ. The problem of type 0 is positive regular. If $\rho > 0$ one no longer has the classical theorem that "short extremals minimize."

Corresponding to a smooth vector function η is a *companion* $\xi = R\eta' + Q\eta$, denoted by ξ^η when convenient. The definition and notation are extended to matrices of m rows and any number of columns. The companion is sometimes more convenient than the derivative.

In addition to the differential equation one considers end conditions $M^r(a^1)$, represented in parametric form by

$$M^r(a^1) : \begin{cases} \eta(a^1) = cu \\ c^T\xi(a^1) = bu \end{cases} \tag{1.4}$$

Here c is an $m \times r$ matrix of constants of rank $r \leqq m$, u is a column of r parameters, and b is an $r \times r$ symmetric matrix of constants. The first set of conditions in (1.4) is called the *end plane* and the second set the *transversality conditions*. The case $r = 0$ of (1.4) is

$$M^o(a^1) : \eta(a^1) = 0 . \tag{1.5}$$

An extremal η which, with a set u of parameters and its companion ξ, satisfies conditions (1.4) will be said to *initiate* at $M^r(a^1)$.

Two extremals η, $\bar{\eta}$ with companions ξ, $\bar{\xi}$ satisfy the identity

$$\eta^T\bar{\xi} - \bar{\eta}^T\xi = \text{const} . \tag{1.6}$$

The extremals are called *conjugate* if the constant is 0. The maximum dimension of a vector space of mutually conjugate solutions is m and the set of extremals that initiate at boundary conditions $M^r(a^1)$ form such a vector space of dimension m.

Conversely, if H is a maximal vector space of mutually conjugate solutions and a^1 is given, there are a unique r depending on a^1 and a set of boundary conditions $M^r(a^1)$ of the form (1.4) at which the elements of H initiate. The representation is not unique. The change of parameter $u = \alpha\hat{u}$, where α is an $r \times r$ nonsingular matrix, replaces (1.4) by

$$M^r(a^1) : \begin{cases} \eta(a^1) = \hat{c}\hat{u} \\ \hat{c}^T\xi(a^1) = \hat{b}\hat{u} \end{cases} \tag{1.7}$$

where $\hat{c} = c\alpha$ and $\hat{b} = \alpha^T b\alpha$. Any two representations of a family H are related by such a change of parameter.

A maximal set of mutually conjugate solutions will be called a *conjugate family*. The *focal points* of a conjugate family are the points at which a nontrivial member of the family vanishes.

If U is an $m \times m$ matrix whose columns are a basis of the family, then the zeros of $|U|$ are the focal points of the family.

Focal points of a family H are to be counted in two ways, of which the first will now be defined while the second is postponed to Section 3. The *order* ν is a function whose value $\nu(e)$ is the order of e as a focal point and is the dimension of the subspace of H of solutions vanishing at e. It is equal to the deficiency in rank of the matrix $U(e)$. Points e with $\nu(e) = 0$ are called *ordinary points*, whereas the focal points are the points e with $\nu(e) > 0$. The point a^1 may be a focal point of the family H presented in (1.4). Its order is $m - r$.

When $r = 0$ in the presentation (1.4) focal points are called conjugate points of a^1. However, by convention the focal point a^1 of order m is not counted as a conjugate point of a^1.

If e is not conjugate to a, there is a unique extremal joining (a,z) to (e,w).

Corresponding to a closed interval $[a^1,a^2]$ there is an $\varepsilon > 0$ such that no two points a, e with $a^1 \leqq a < e \leqq a^2$ and $e - a < \varepsilon$ are conjugate.

2. CONDITIONS FOR ISOLATED FOCAL POINTS

Focal points in a problem of type ρ with $0 < \rho < m$ need not be isolated. In an Appendix there is an example with $m = 2$ and $\rho = 1$ in which the point $a^1 = -1$ has no conjugate point on $(-1,0)$ but in which each point of $[0,1]$ is conjugate to -1.

With end conditions $M^r(a^1)$ there will be associated two quadratic forms. The *primary form* depends only on the end plane. It is the form $z^TR(a^1)z$ cut down to the end plane, namely,

$$\kappa^1(u) = u^Tc^TR(a^1)cu \; . \tag{2.1}$$

When $r = 0$, the form is trivial and by convention is nonsingular. The *secondary form* is defined on the characteristic plane M_0^ν of the primary form. It is

$$\kappa^2(u) = u^T[b + c^TS(a^1)c]u|M_0^\nu \tag{2.2}$$

with $S = \frac{1}{2}[R' - (Q + Q^T)]$. Although the two forms depend on the representation of the end conditions, a change of representation yields respectively equivalent forms.

THEOREM 2.1 If the primary form of the end plane in boundary conditions $M^r(a^1)$ is nonsingular or if the primary form is singular but the secondary form is nonsingular, then there is a deleted interval of a^1 containing no focal point of the conjugate family H initiating at end conditions $M^r(a^1)$.

For notational convenience, suppose $a^1 = 0$. The proof in case $r = 0$ is the classical one, a result already noted in Section 1. When $r > 0$, there is a unique extremal η joining $(0,cu)$ to $(t,0)$ when $t > 0$ and t is near 0. It is convenient to use that basis $(v \;\; w)$ of solutions of the differential equation for which

$$\left.\begin{pmatrix} v & w \\ v' & w' \end{pmatrix}\right|_{x=0} = 1 \tag{2.3}$$

and to calculate $w''(0) = -R^{-1}(0)[R'(0) + Q(0) - Q^T(0)]$. Then

$$\eta = [v - ww^{-1}(t)v(t)]cu \; . \tag{2.4}$$

The point t is a focal point if and only if η is a nontrivial member of H, that is, $u \neq 0$ and $bu - c^T\xi(0) = 0$. Here ξ is the companion of η, that is, $\xi = [\xi^v - \xi^w w^{-1}(t)v(t)]cu$. This is the case precisely when the quadratic form

$$\begin{aligned} \theta(u) &= u^T bu - u^T c^T \xi(0) \\ &= u^T\{b + c^T[-Q(0) + R(0)w^{-1}(t)v(t)]c\}u \end{aligned} \tag{2.5}$$

is singular. If one writes

$$\begin{aligned} v(x) &= 1 + xA(x) \qquad & A(0) &= v'(0) = 0 \\ w(x) &= xB(x) & B(0) &= w'(0) = 1 \\ & & B'(0) &= \tfrac{1}{2}w''(0) \end{aligned} \tag{2.6}$$

then $\theta(u) = \frac{1}{t}\gamma(u)$ where

$$\begin{aligned} \gamma(u) = u^T\{&c^T R(0)B^{-1}(t)c \\ &+ t[b + c^T(-Q(0) + R(0)B^{-1}(t)A(t))c]\}u . \end{aligned} \tag{2.7}$$

If $t = 0$, then $\gamma(u) = \kappa^1(u)$. If the latter is nonsingular, then $\theta(u)$ is nonsingular for t in a deleted interval about 0 and the theorem is proved.

In case $\kappa^1(u)$ is singular, with characteristic plane M_0^ν, one calculates

$$\frac{d}{dt}\gamma(u)\Big|_{t=0} = u^T bu + u^T c^T S(0)cu \tag{2.8}$$

to see that

$$\frac{d}{dt}\gamma(u)\Big|_{t=0}\Big| M_0^\nu = \kappa^2(u) . \tag{2.9}$$

It follows from a theorem of the author [5] that $\gamma(u)$ is nonsingular for t in a deleted neighborhood of 0 if $\kappa^2(u)$ is nonsingular, so that the proof is complete.

There will be a subsequent reference to $\theta(u)$ when $t > 0$, for which the following is needed.

COROLLARY 2.1 If κ^1 is nonsingular, the index of θ for $t > 0$ and sufficiently near 0 equals the index of κ^1. If κ^1 is singular but κ^2 is nonsingular, then the index of θ for $t > 0$ and sufficiently near 0 is the sum of the indices of κ^1 and κ^2.

This follows from the same reference [5].

The criterion of Theorem 2.1 is generally applicable. Given a family H, any point a can be tested by the sufficient condition that it be isolated as a focal point by representing H as initiating at a and examining the corresponding κ^1 and κ^2. The result corresponding to κ^1 nonsingular is as follows.

COROLLARY 2.2 The focal point at a of the conjugate family with basis y is isolated if $y(a)^T R(a) y(a)$ has the same rank as $y(a)$.

The result corresponding to κ^1 singular and κ^2 nonsingular is as follows.

COROLLARY 2.3 Let y be a basis of the conjugate family H such that $y(a) = (c_1\ c_2\ 0)$ where $(c_1\ c_2)$ has r columns and rank r and c_2 has ν columns and

$$y(a)^T R(a) y(a) = \begin{pmatrix} c_1^T R(a) c_1 & 0 \\ 0 & 0 \end{pmatrix} \tag{2.10}$$

has rank $r - \nu$. Let $b = \begin{pmatrix} b_{11} & b_{12} \\ b_{21} & b_{22} \end{pmatrix}$ with $r - \nu$ and ν rows and columns. Then a is an isolated focal point of H if the $\nu \times \nu$ matrix

$$(2.11) \qquad \tfrac{1}{2}c_2^T[R'(a) - (Q(a) + Q^T(a))]c_2 + b_{22}$$

is nonsingular.

Corollary 2.2 is the special case $\nu = 0$ of Corollary 2.3, again with the convention that the 0×0 matrix is nonsingular.

A conjugate family will be called *regular* if its focal point are isolated and *completely irregular* if every point is a focal point. The following condition is of interest.

CONDITION F Each conjugate family of the differential equation is either regular or completely regular.

Consider the following hypothesis.

HYPOTHESIS A The elements of R, Q, P are analytic functions.

THEOREM 2.2 Hypothesis A implies Condition F.

The theorem is transparent because the determinants whose zeros determine focal points are analytic functions.

A function Δ of one variable called the *focal discriminant* will be defined. One then considers the following hypothesis.

HYPOTHESIS B The coefficient matrices R, Q, P have continuous derivatives of order $p = \binom{2m}{m}$ and the focal discriminant Δ does not vanish.

THEOREM 2.3 Hypothesis B implies Condition F.

A sketch of the proof involving the use of the focal discriminant will be given first and the definition and outline of the calculation will be described afterwards. It will be shown that there is a linear homogeneous differential equation

$$\Delta y^{(p)} + \sum_{i=1}^{p} \Delta_i y^{(p-i)} = 0 \tag{2.12}$$

that is satisfied by every determinant $D = |U|$, where U is an $m \times m$ matrix whose columns are solutions of the differential equation. The function Δ is the focal discriminant. When it has been defined, the truth of the theorem is apparent.

To define and calculate Δ, one rewrites the given differential equation in the form

$$\begin{pmatrix} \eta \\ \eta' \end{pmatrix}' = B \begin{pmatrix} \eta \\ \eta' \end{pmatrix} \tag{2.13}$$

where B is a $2m \times 2m$ matrix computed from P, Q, R. Specifically

$$B = R^{-1} \begin{pmatrix} 0 & R \\ -Q'+P & -R'-Q+Q^T \end{pmatrix} . \tag{2.14}$$

Let $W = \begin{pmatrix} U \\ U' \end{pmatrix}$, with U as just defined. Let U_i, $i = 1,\dots,p$ be the $m \times m$ submatrices of W, with $U_1 = U$, and let $D_i = |U_i|$, so that $D_1 = D = |U|$. By differentiating a row at a time and using (2.13), one finds that

$$D_i' = \Sigma_j \, c_{ij} D_j . \tag{2.15}$$

The coefficients c_{ij} are independent of U. They are rational functions in which the numerators are polynomials in elements of P, Q, R, Q', R' and the denominator is a polynomial in elements of R, namely $|R|$. Now beginning with D_1' and repeatedly differentiating and substituting from (2.15), one finds

$$D_1^{(h)} = \Sigma_j c_j^h D_j \quad , \quad h,j = 1,\ldots,p \tag{2.16}$$

where coefficients c_j^h are rational functions whose numerators are polynomials in elements of P , Q , R and their derivatives of order through h and whose denominators are powers of $|R|$. Again, the coefficients c_j^h are independent of U . The equations (2.16) are multiplied by appropriate cofactors and added to eliminate $D_2,\ldots,D_p$ and the result is multiplied by the highest power of $|R|$ then appearing in any denominator. This yields the equation (2.12) and defines Δ as a polynomial in the elements of P , Q , R and their derivatives of order at most p .

Inasmuch as Hypothesis A or B implies Condition F, it is reasonable to assume that Condition F holds. This will be done when convenient but it is not a permanent assumption. In discussing a single conjugate family or in comparing two conjugate families, it may be sufficient to assume that each is completely regular without assuming Condition F.

Theorem 2.1 has the following consequences.

COROLLARY 2.4 In the presence of Condition F, the conjugate points of a point a or the focal points of any end conditions $M^r(a)$ with primary form κ^1 or secondary form κ^2 nonsingular are isolated. If any focal point of a conjugate family is isolated then all are.

3. THE SIGNATURE OF AN ISOLATED FOCAL POINT

Focal points will be counted in a second way, with weight equal to the negative of a signature or a differential signature to be defined. This agrees with counting by weight equal to order when $\rho = 0$, since then the two signatures agree and their common value is the negative of the order.

The two signatures will be defined for a point e which is an isolated focal point of a conjugate family H. The definitions and theorems seem to be within reach for focal points that are not isolated but the interest may not justify the complexity.

The motivation for the definitions appears in Section 4. The definitions and some calculations are presented first in order to emphasize their local character.

Suppose that $a < e < f$, that e is the only focal point of H on $[a,f]$, and that a has no conjugate point on $(a,f]$. Represent H as initiating at $M^r(a^1)$ with $a^1 < a$. Let η^1 be the member of H through (a,z) and let $\eta^2(\cdot,t)$ be the extremal joining (a,z) to $(t,0)$ with $a < t \leqq f$. Let

$$\overline{Q}(z,t) = u^T bu + \int_{a^1}^{t} 2\omega(\eta,\eta')dx \tag{3.1}$$

where $\eta = \eta^1$ on $[a^1,a]$ and $\eta = \eta^2(\cdot,t)$ on $[a,t]$. Then

$$\overline{Q}(z,t) = \eta^1(a)^T\xi^1(a) - \eta^2(a,t)^T\xi^2(a,t) \tag{3.2}$$

where ξ^1 and ξ^2 are companions of η^1 and η^2. Thus $\overline{Q}$ is defined from H on an arbitrarily small neighborhood of e as a quadratic form in z, dependent on a parameter t. It is exactly the form $\theta(u)$ of Section 2, with $r = m$ and $c = 1$. It is singular if and only if there exists a z for which $\xi^1(a) = \xi^2(a,t)$, that is, $\eta^1 = \eta^2$, meaning that t is a focal point of H. Thus it is nonsingular with index $h_{\overline{Q}}(t)$ independent of t, for $a < t < e$ and for $e < t \leqq f$.

LEMMA 3.1 The index $h_{\overline{Q}}(t) = \rho$ for $a < t < e$.

This is an instance of Corollary 2.1. Direct calculation is slightly simpler than the full calculation of Theorem 2.1.

The *signature* of e as a focal point of H is defined as

$$\sigma(e) = h_{\overline{Q}}(a+) - h_{\overline{Q}}(f) = \rho - h_{\overline{Q}}(f) . \tag{3.3}$$

The change in index of $\overline{Q}$ as t increases through e is thus $-\sigma(e)$. The apparently gratuitous minus sign was not absorbed in the definition in order to make the definition agree in the next paragraph with that of a classical signature.

Observe that $|\sigma(e)| \leqq \nu(e)$.

In the same context, let H_e denote the subspace of elements of H that vanish at e and let

$$q^e = \overline{Q}_t(\cdot,e)|H_e . \tag{3.4}$$

This is a quadratic function in $\nu(e)$ variables with negative index $h^o(e)$, nullity $\nu^o(e)$, and positive index $1^o(e)$, for which $\nu(e) = h^o(e) + \nu^o(e) + 1^o(e)$. Its signature is

$$\sigma^o(e) = 1^o(e) - h^o(e) \tag{3.5}$$

and will be called the *differential signature* of e as a focal point of H. The change in index of $\overline{Q}(\cdot,t)$ at $t = e$, already established as $-\sigma(e)$, lies between $-\sigma^o(e) - \nu^o(e)$ and $-\sigma^o(e) + \nu^o(e)$. In particular, if q^e is nondegenerate, the signature and differential signature agree. See [5].

LEMMA 3.2 When the slopes $p = \eta'(e)$ for $\eta \in H_e$ are introduced as variables

$$q^e(z) = -p^T R(e) p . \tag{3.6}$$

One begins with

(3.7) $$\overline{Q}(z,t) = \eta^1(a)^T\xi^1(a) + \int_a^t 2\omega(\eta(x,t),\eta_x(x,t))dx$$

where $\eta(x,t)$ has been written in place of $\eta^2(x,t)$. Then

$$\overline{Q}_t(z,t) = 2\omega[\eta(t,t),\eta_x(t,t)]$$

$$+ \int_a^t 2[\omega_\eta\eta_t + \omega_\zeta\eta_{xt}]dx$$

(3.8) $$= \eta_x^T R\eta_x|^{x=t} + 2\eta_t(x,t)\omega_\zeta(\eta(x,t),\eta_x(x,t))|_{x=a}^{x=t} .$$

Since $\eta(a,t) = z$, it follows that $\eta_t(a,t) = 0$. Since $\eta(t,t) = 0$ it follows that $\eta_x(t,t) + \eta_t(t,t) = 0$. Thus

(3.9) $$\overline{Q}_t(z,t) = -\eta_x R\eta_x|^{x=t} .$$

The slopes p at t can be introduced as independent variables in $\overline{Q}_t$ in place of z. The form of the change of variable is

(3.10) $$p = G(t)z , \quad G(t) = v'(t) - w'(t)w^{-1}(t)v(t)$$

where v, w is a basis of solutions for which $\begin{pmatrix} v & w \\ v' & w' \end{pmatrix}\Big|_{x=a} = 1$, but it is not needed. Then $\overline{Q}(z,t) = -p^T R(t)p$ and

(3.11) $$q^e(z) = -p^T R(t)p , \quad p = \eta'(e) \text{ and } \eta \in H_e .$$

4. THE ADJUSTED INDEX

The adjusted index is defined corresponding to the differential equation (1.1) and general self-adjoint boundary conditions

$$(4.1)\qquad M^r(a^1,a^2) : \begin{cases} \eta(a^1) = c^1u \\ \eta(a^2) = c^2u \\ c^{2T}\xi(a^2) - c^{1T}\xi(a^1) + bu = 0 \end{cases}$$

in which $0 \leqq r \leqq 2m$, c^1 and c^2 are $m \times r$ matrices such that $\begin{pmatrix} c^1 \\ c^2 \end{pmatrix}$ has rank r , b is an $r \times r$ symmetric matrix of constants, ξ is the companion of η , and $a^1 < a^2$.

The definition begins with the broken extremal construction of Morse. See [3, Ch. III, Sect. 1] or [4, Sect. 13]. Let $\varepsilon > 0$ be so small that no two points of $[a^1,a^2]$ at mutual distance less than ε are conjugate. Suppose that π is a partition

$$(4.2)\qquad a^1 = a_0 < a_1 < \cdots < a_{N+1} = a^2$$

with $a_{j+1} - a_j < \varepsilon$. Suppose that $Z = (u,z^1,\ldots,z^N)$ is a set of $r + Nm$ variables, z^j being an m-vector. As a temporary notation let $\bar{\eta}(\cdot;e^1,w^1;e^2,w^2)$ be the extremal with

$$(4.3)\qquad \bar{\eta}(e^s;e^1,w^1;e^2,w^2) = w^s , \quad s = 1,2$$

and let

$$(4.4)\qquad \eta(x;\pi,Z) = \begin{cases} \bar{\eta}(x;a_0,c^1u;a_1,z^1) & , a_0 \leqq x \leqq a_1 \\ \bar{\eta}(x;a_j,z^j;a_{j+1},z^{j+1}) & , a_j \leqq x \leqq a_{j+1} \\ & \quad j = 1,\ldots,N-1 \\ \eta(x;a_N,z^N;a_{N+1},c^2u) & , a_N \leqq x \leqq a_{N+1} \end{cases} .$$

This is the broken extremal with corners at (a_j,z^j) and satisfying the end conditions of (4.1). Now let

$$(4.5)\qquad Q(Z;\pi) = u^Tbu + \int_{a^1}^{a^2} 2\omega(\eta(x;\pi,Z),\eta_x(x;\pi,Z)dx .$$

is a quadratic form in Z. Calculation shows that it is degenerate if and only if the Euler equation (1.1) has a solution satisfying the boundary conditions (4.1) and that the nullity ν of Q is the dimension of the vector space of such solutions. Thus the index i^N and nullity of Q are independent of the location of the intermediate points on the partition π.

With a^2 sufficiently close to a^1 the value $N = 0$ is admissible in the definition above.

In the classical case in which $\rho = 0$, one shows that i^N is, in fact, independent of N, either directly or by showing that it is equal to the index of the quadratic function

$$I(\eta,u) = u^T bu + \int_{a^1}^{a^2} 2\omega(\eta,\eta')dx \tag{4.6}$$

on the class of curves of D' satisfying end conditions

$$\eta(a^1) = c^1 u , \quad \eta(a^2) = c^2 u . \tag{4.7}$$

When $\rho \neq 0$, the index i^N does depend on N and the index of I is $+\infty$. Accordingly, an adjusted index is attached to the integral I subject to the end conditions (4.7) or to the differential equation (1.1) subject to the boundary conditions $M^r(a^1,a^2)$. The *adjusted index* is $\hat{i} = i^N - N\rho$. It will be shown to be independent of N, with N sufficiently large that i^N is defined. There are preliminary steps to be taken.

Suppose that the partition $\pi = \{a_0,a_1,a_2\}$ is admissible and that η is the extremal with possibly a single corner at a_1, satisfying $\eta(a_o) = 0$, $\eta(a_1) = z$, $\eta(a_2) = 0$. Let

$$\begin{aligned} Q^1(z) &= \int_{a^1}^{a^2} 2\omega(\eta,\eta')dx \\ &= z^T[\xi(a_2-) - \xi(a_1+)]z . \end{aligned} \tag{4.8}$$

LEMMA 4.1 With $a_2 - a_0$ sufficiently small, the partition π is admissible for all a_1 on (a_0, a_2) and the index of Q_1 is ρ.

Here a_2 is so close to a_0 that there is no conjugate point of a_0 on $(a_0, a_2]$. The result is then a special case of Lemma 2.1.

There is a lemma of Morse on quadratic forms that will be restated in terms of its application here. See [3, Ch. III, Sect. 7] or [4, Sect. 25]. Let J and J' denote complementary subsets of $\{0,1,\ldots,N\}$.

LEMMA 4.2 Let Q_1 be the form obtained from Q by setting $z^j = 0$ for $j \in J$. This is interpreted as $u = 0$ if $0 \in J$. Suppose that Q_1 is nonsingular. Let Q_2 be the form Q with the variables z^j, $j \in J'$, eliminated by the condition that there be no corner at a_j. This is interpreted as satisfying the conditions $M^r(a^1, a^2)^j$ if $0 \in J'$. Then $i_Q = i_{Q_1} + i_{Q_2}$ and $\nu_Q = \nu_{Q_2}$.

THEOREM 4.1 The adjusted index $\hat{i}$ of I is independent of N.

For proof, let the partition π define Q^N and insert one vertex a with $a_k < a < a_{k+1}$ to obtain the partition π' defining Q^{N+1}. Apply Lemma 4.2 to Q^{N+1}, setting the variables at each vertex of π equal to 0. Lemmas 4.1 and 4.2 show that $i_{Q^{N+1}} = i_{Q^N} + \rho$ so that

$$(4.9) \qquad \hat{i}_{Q^{N+1}} = i_{Q^{N+1}} - (N+1)\rho = i_{Q^N} - N\rho = \hat{i}_{Q^N}.$$

This completes the proof.

. COUNTING FOCAL POINTS

The point t is a focal point of the conjugate family H nitiating at $M^r(a^1)$ exactly when the quadratic function

$$I(\eta) = u^T bu + \int_{a^1}^{t} 2\omega(\eta,\eta')dx \tag{5.1}$$

ith end conditions $\eta(a^1) = cu$ and $\eta(t) = 0$ is singular. ocal points of H will be counted by use of the adjusted index (t) of $I(\eta)$, with weights of focal points equal to the signa- ure or the differential signature. In the calculation, the ndex $i^N(t)$ of $I(\eta)$ corresponding to a partition with N ntermediate vertices will appear.

THEOREM 5.1 The adjusted index of the regular conjugate amily H determined by end conditions $M^r(a^1)$ is given by

$$\hat{i}(a^2-) = \hat{i}(a^1+) - \Sigma\sigma(b) , \quad a^1 < b < a^2 . \tag{5.1}$$

loreover

$$|\hat{i}(a^2-) - \hat{i}(a^1+) + \Sigma\sigma^0(b)| \leqq \Sigma\nu^0(b) . \tag{5.2}$$

The quantity $\hat{i}(a^1+)$ is the index of the form $\theta(u)$ intro- luced in Theorem 2.1 . If the form $\kappa^1(u)$ or $\kappa^2(u)$ is non- singular, it is evaluated in Corollary 2.1. The extreme cases ccur with the end conditions $M^0(a^1)$, when $\hat{i}(a^1+) = 0$, and $M^m(a^1)$, when $\hat{i}(a^1+) = \rho$.

The proof lies in considering the adjusted index $\hat{i}(t)$ of (η) as t increases from a_1+ to a_2 . More specifically, uppose e is a focal point of H with $a^1 < e < a^2$ and π is a partition with $a_N < e < a_{N+1}$ such that H has no other focal point on $[a_N, a_{N+1}]$. Apply Lemma 4.2 with $J = \{N\}$. Thus

$$i^{N+1}(a_{N+1}) = i^N(a_N) + h_{\overline{Q}}(a_{N+1}) . \tag{5.3}$$

The last term is the index of I on the m dimensional space of broken extremals with a single corner, where the segment on $[a^1, a_N]$ is the member of H through (a_N, z) and the segment on $[a_N, a_{N+1}]$ is the extremal joining (a_N, z) to $(a_{N+1}, 0)$. The quadratic form is the form $\overline{Q}$ of Section 3, whose index is $-\sigma(e)$. Finite induction completes the proof of (5.1) and the relation $|\sigma(e) - \sigma^0(e)| \leq \nu^0(e)$ noted in Section 3 yields (5.2)

6. COMPARISONS

Theorems relating the values of the adjusted index associated with two sets of end conditions on the same interval are readily established. One general instance is offered. The problem B consisting of the differential equation (1.1) with general boundary conditions $M^r(a^1, a^2)$ will be compared with the problem B_0 consisting of the same equation with the boundary conditions

$$M^0(a^1, a^2) : \eta(a^1) = 0 , \ \eta(a^2) = 0 . \tag{6.1}$$

Respective adjusted indices will be denoted by $\hat{i}$ and $\hat{i}_0$.

Only the case that a^2 is not conjugate to a^1 is presented. However, there is no assumption of regularity. The form

$$Q_0(u) = u^T bu + \int_{a^1}^{a^2} 2\omega(\eta, \eta')dx \tag{6.2}$$

valuated on the unique extremal joining (a^1, c^1u) to (a^2, c^2u) s well-defined. Its index and nullity will be denoted by h_o nd ν_o .

THEOREM 6.1 The adjusted index of problem B is $= \hat{i}_o + h_o$. The number of solutions of the problem B is o .

The theorem follows from Lemma 4.2 applied to the form $(Z;\pi)$ of (4.5) using $J = \{0\}$.

As a special case one may compare the count of focal points f end conditions $M^r(a^1)$ on $[a^1, a^2)$ with the number of conugate points of a^1 on (a^1, a^2) provided the two conjugate amilies are completely regular and the count is weighted by the ignatures.

. CHARACTERISTIC ROOTS

One may consider the boundary value problem

$$
\text{7.1)} \qquad B(\lambda) : \begin{cases} \dfrac{d}{dx}\omega_\zeta - \omega_\eta + \lambda\eta = 0 \\ \eta(a^1) = c^1u , \quad \eta(a^2) = c^2u \\ c^{2T}\xi(a^2) - c^{1T}\xi(a^1) + bu = 0 \end{cases} .
$$

t corresponds to an integrand $\tilde{\omega}(\eta,\zeta,\lambda) = \omega(\eta,\zeta) - \lambda\eta^T\eta$ for hich the quadratic form $\tilde{Q}(Z,\lambda;\pi)$ depending on the parameter can be set up. The adjusted index $\hat{i}(\lambda)$ depends on λ . One ay consider a bounded interval in λ with a single sufficiently ine partition π or adjust the mesh of the partition as needed. hen the following theorem is readily established.

THEOREM 7.1 If $\mu^1 < \mu^2$ and neither is a characteristic root of the problem $B(\lambda)$, then the number of characteristic roots of $B(\lambda)$ on $[\mu^1, \mu^2]$ is $\hat{i}(\mu^2) - \hat{i}(\mu^1)$.

REFERENCES

1. Bliss, G. A. A boundary value problem for a system of ordinary differential equations, *Trans. Amer. Math. Soc.* *28* (1926), pp. 561-584.

2. Hu, K.-S. The problem of Bolza and its accessory boundary value problem, "Contributions to the Calculus of Variations." pp. 361-443. University of Chicago, 1931-32.

3. Morse, M. The calculus of variations in the large, *Amer. Math. Soc. Coll. Publ. XVIII*, 1934.

4. Morse, M. "Variational Analysis: Critical Extremals and Sturmian Extensions." Wiley, New York, 1973.

5. Pitcher, E. The variation in index of a quadratic function depending on a parameter, *Bull. Amer. Math. Soc.* *65*(1959), pp. 355-357.

6. Reid, W. T. A system of ordinary linear differential equations with two-point boundary conditions, *Trans. Amer. Math. Soc.* *44*(1938), pp. 508-521.

7. Reid, W. T. "Ordinary Differential Equations." Wiley, New York, 1971.

ADDENDUM

When an extended abstract of this paper was presented, W. T. Reid pointed out an alternate route. He supplied references [1]; [2; particularly pp. 361-413]; [6, particularly Theorem 6.3]; [7, particularly Problem 8 on pp. 396-397 and Ch. IV, Sect. 6].

ne writer is grateful for the reminder of the method and for the eferences.

PPENDIX

The example mentioned in Section 2 is the following. Let

$$g(x) = 6x^3 + 6x^2 - 1 \qquad -1 \leqq x \leqq 0$$
$$= -1 \qquad 0 \leqq x \leqq 1 \quad .$$

nen $g \in C'$ and $g(x) \neq 0$. Let

$$R(x) = \frac{1}{g(x)} \begin{pmatrix} 3x^2+2x & 3x^2-1 \\ 3x^2-1 & 3x^2 \end{pmatrix} \qquad -1 \leqq x \leqq 0$$

$$= \frac{1}{g(x)} \begin{pmatrix} 2x & -1 \\ -1 & 0 \end{pmatrix} \qquad 0 \leqq x \leqq 1 \quad .$$

et $2\omega(\eta,\zeta) = \zeta^T R \zeta$, so that the differential equation under onsideration is

$$\frac{d}{dx}(R\eta') = 0 \; .$$

matrix whose columns are a basis for the conjugate family atisfying the end condition $\eta(-1) = 0$ is

$$U(x) = \begin{pmatrix} x^3+1 & -x^3+x \\ -x^3+x & x^3+x^2 \end{pmatrix} \qquad -1 \leqq x \leqq 0$$

$$= \begin{pmatrix} 1 & x \\ x & x^2 \end{pmatrix} \qquad 0 \leqq x \leqq 1 \quad .$$

It is readily calculated that $RU' = 1$. One sees that

$$\begin{aligned} |U(x)| &= x^3(x+1)^2 & -1 \leqq x \leqq 0 \\ &= 0 & 0 \leqq x \leqq 1 \ . \end{aligned}$$

This is a system with $m = 2$ of type $\rho = 1$. The set of conjugate points of -1 on $(-1,1]$ is $[0,1]$.

POSITIVE FUNCTIONALS AND OSCILLATION CRITERIA FOR DIFFERENTIAL SYSTEMS

Garret J. Etgen

University of Houston
Houston, Texas

Roger T. Lewis

University of Alabama
Birmingham, Alabama

1. INTRODUCTION

Let H be a Hilbert space, let $\mathcal{B} = \mathcal{B}(H,H)$ be the B^*-algebra of bounded linear operators from H to H with the uniform operator topology, and let S be the subset of $\mathcal{B}$ consisting of the selfadjoint operators. This article is concerned with the second order selfadjoint differential equation

$$L[Y] = [P(x)Y']' + Q(x)Y = 0 \tag{1}$$

ISBN 0-12-632250-3

on $R^+ = [0,\infty)$, where $P,Q : R^+ \to S$ are continuous with $P(x)$ positive definite for all $x \in R^+$. Appropriate discussions of the concepts of integration and differentiation of B-valued functions, as well as treatments of the existence and uniqueness of solutions $Y : R^+ \to B$ of (1) can be found in a variety of texts. See, for example, E. Hille [19, Chapters 6 and 9]. In particular, it is well-known that when suitable initial conditions are specified for (1), then the resulting initial value problem has a unique solution.

In this paper we discuss the behavior of solutions of (1) with particular emphasis on the oscillation of solutions. Studies of the behavior of solutions of second order equations in Banach spaces have been made by several authors, including Hille [19, Chapter 9], T. L. Hayden and H. C. Howard [17], G. J. Etgen and J. F. Pawlowski [12,13], E. S. Noussair [33], and C. M. Williams [47]. The doctoral dissertation of Williams gives a complete treatment of the basic theory of equation (1), and includes existence and uniqueness of solutions, the relationship between (1) and the Riccati equation, nonoscillation, oscillation and disconjugacy.

It is important to note that if $H = R_n$, Euclidean n-space, then B is the B^*-algebra of $n \times n$ matrices and equation (1) is the familiar second order selfadjoint matrix differential equation which has been investigated in great detail by a large number of authors. In this regard, we refer to the texts by F. V. Atkinson [5], P. Hartman [15], E. Hille [19], M. Morse [31], W. T. Reid [36,37] and C. A. Swanson [39], as well as to the research papers of C. D. Ahlbrandt, W. Allegretto and L. Erbe, A. Coppel, G. J. Etgen, P. Hartman, H. C. Howard, K. Kreith, R. T. Lewis, E. S. Noussair, W. T. Reid, C. A. Swanson, E. C. Tomastik, and V. A. Yakubovic. While there are a variety of reasons for considering the behavior of solutions of (1) in either the matrix or general B^*-algebra case, much of the motivation for such investigations comes from the tremendous amount of research devoted to the second order scalar equation

$$(2) \qquad [p(x)y']' + q(x)y = 0 ,$$

where p and q are continuous, real-valued functions on R^+ with $p(x) > 0$ for all $x \in R^+$. Research on this equation dates to the work of J. Liouville and C. Sturm in the 1830's, and includes the work of authors far too numerous to mention specifically here.

2. BASIC DEFINITIONS AND EXAMPLES

Throughout this paper we shall assume that H is a Hilbert space over the reals R, with the inner product on H denoted by $<,>$ and norm $\| \; \| = <,>^{1/2}$. It will be apparent that the methods and results in the paper apply equally as well when H is a Hilbert space over the field of complex numbers, but we restrict our attention to real Hilbert spaces because most of the work dealing with the matrix version of (1) and the scalar equation (2) has been done over the reals.

We shall assume that the B^*-algebra B of bounded linear operators from H to H is topologized by the operator norm

$$\|A\| = \sup_{\|\alpha\|=1} <A\alpha,\alpha> .$$

In the case where $H = R_n$, the space of ordered n-tuples of real numbers, B is the B^*-algebra of $n \times n$ matrices with the $*$-operation being "transpose." The case $H = R_n$ will be referred to as the *finite dimensional case,* except when $n = 1$. The case $n = 1$, i.e., $H = R_1 = R$ and (1) $\equiv$ (2) will be called the *scalar case*. The symbol I is used for the identity element of $\mathcal{B}$. The symbol 0 is used indiscriminately for the zero element, with the appropriate interpretation being clear from the context. If $A \in S$, the selfadjoint elements of $\mathcal{B}$, then the notation $A > 0$ ($A \geq 0$) is used to signify that A is positive (nonnegative) definite.

Let $Y = Y(x)$ be a solution of equation (1). Then it is easy to verify by differentiation that

$$Y^*[PY'] - [PY']^*Y \equiv C \quad \text{(constant)}$$

on R^+.

DEFINITION 2.1 A solution $Y = Y(x)$ of equation (1) is *conjoined* (or *prepared*) if

$$Y^*[PY'] - [PY']^*Y \equiv 0$$

on R^+.

The term "conjoined" has its origins in the Calculus of Variations, and for amplifications of this concept the reader is referred to M. Morse [31], and to W. T. Reid [36]. Conjoined solutions of (1) can be obtained simply by choosing conjoined initial values. In fact, it is easy to show that Y is a conjoined solution of (1) if and only if there is at least one point $a \in R^+$ such that

$$Y^*(a)[P(a)Y'(a)] = [P(a)Y'(a)]^*Y(a) .$$

The following example, given by Noussair and Swanson [32], shows that the conjoined hypothesis on solutions of (1) is needed in order to obtain an analog of the classical theory of oscillation of the scalar equation (2).

EXAMPLE 2.1 Let $H = R_2$, and consider the differential equation

$$(IY')' + IY \equiv Y'' + Y = 0 \quad \text{on} \quad R^+ .$$

As an analog of the scalar equation $y'' + y = 0$, all of whose solutions are of the form $y(x) = a \cdot \sin(x + b)$, a, b constant, one would want all of the solutions of the given equation to oscillate. (In this case a solution Y is oscillatory if and only if the determinant of Y, $\det Y$, has infinitely many zeros on R^+.) The matrix-valued function

$$Y(x) = \begin{pmatrix} \cos x & -\sin x \\ \sin x & \cos x \end{pmatrix}$$

is a solution of the equation which is not conjoined, and which does not oscillate since $\det Y \equiv 1$ on R^+. It can be shown that every conjoined solution of the equation is oscillatory.

DEFINITION 2.2 A solution $Y = Y(x)$ of equation (1) is *nonsingular at* $x = a$, $a \in R^+$, if

(i) the range of $Y(a) : H \to H$ is H, and

(ii) $Y(a)$ has a bounded inverse.

If either of these conditions fails to hold at $x = a$, then Y is *singular at* $x = a$. The solution Y has an *algebraic singularity at* $x = a$ if $Y(a)$ is not one-to-one.

In the finite dimensional case it is clear that the only singularities of a solution Y of (1) are algebraic singularities, and that Y is singular at $x = a$ if and only if $\det[Y(a)] = 0$. In the general B^*-algebra case, conditions (i) and (ii) in Definition 2.2 are equivalent to the statement $Y^{-1}(a) \in B$.

Definitions 2.1 and 2.2 are those used by Hille [19], Hayden and Howard [17], Etgen and Pawlowski [12,13], and Williams [47]. E. S. Noussair [33] has introduced slightly different versions of the terms "prepared" and "nonsingular." In particular, he defines:

DEFINITION 2.2' A solution $Y = Y(x)$ of equation (1) is nonsingular at $x = a$ if $Y(a)$ has a bounded inverse.

DEFINITION 2.1' A solution $Y = Y(x)$ of equation (1) is prepared if

(i) $Y^*[PY'] \equiv [PY']^*Y$ on R^+,

(ii) there is a constant vector $\alpha \in H$, $\alpha \neq 0$, such that α is in the range of $Y(a)$ whenever $Y(a)$ is nonsingular.

Note that in Noussair's definition of nonsingularity at $x = a$ it is not required that $Y(a)$ be onto. Noussair gives the following example to illustrate the distinction between his definitions and Definitions 2.1 and 2.2.

EXAMPLE 2.2 Let $H = \ell_2$, and consider the initial value problem

$$Y'' + Y = 0 \text{ on } R^+, \quad Y(0) = 0, \quad Y'(0) = T,$$

where T is the "right shift" operator on ℓ_2, i.e., $T(a_1,a_2,a_3,\ldots) = (0,a_1,a_2,a_3,\ldots)$. The solution of this initial value problem is $Y(x) = (\sin x)T$. Since T is not onto, this solution is identically singular on R^+ according to Definition 2.2. However, as shown by Noussair, if Definitions 2.1' and 2.2' are used, then Y is a prepared solution of the equation which is "nonsingular" whenever $\sin x \neq 0$.

It will be apparent in the work which follows that the methods and results of this paper can be applied regardless of which definitions of "conjoined" and "nonsingular" are used. For convenience in the presentation, we shall use Definitions 2.1 and 2.2 throughout the remainder of the paper.

DEFINITION 2.3 A solution $Y = Y(x)$ of equation (1) is *nontrivial* if there is at least one point $a \in R^+$ such that $Y(a)$ is nonsingular.

In the finite dimensional case it is well-known that a solution Y of (1) is nontrivial if and only if $Y^*Y + [PY']^*[PY'] > 0$ on R^+, and that a nontrivial solution has at most a finite number of singular points on any compact subset of R^+. These properties do not carry over to the general B^*-algebra case. Hayden and Howard [17] have shown that while the set of singularities of a nontrivial solution Y of (1) is a closed set, it is possible for the set of singularities to have a finite limit point. It is easy to show that the condition $Y^*Y + [PY']^*[PY'] > 0$ is necessary for Y to be nontrivial, but in the general B^*-algebra case it is not sufficient as the next example shows.

EXAMPLE 2.3 Let $H = \ell_2$, and consider the differential equation

$$Y'' + AY = 0 \quad \text{on} \quad R^+ ,$$

where A is the infinite diagonal matrix $A = \pi^2 \text{diag}[1,1/4, 1/9,\ldots]$. The function $Y : R^+ \to B$ given by $Y(x) = \text{diag}[\sin \pi x, \sin \pi x/2, \sin \pi x/3,\ldots]$ is a solution with the property $Y^*Y + [PY']^*[PY'] > 0$ on R^+. However, for each fixed $x \in R^+$, x not an integer, Y is one-to-one but does not have a bounded inverse because 0 is in the spectrum of $Y(x)$. Clearly, Y has an algebraic singularity at each integer $n \in R^+$. Thus Y is identically singular on R^+ .

We turn now to the question of the oscillation of solutions of (1). For the remainder of the paper we shall assume that the term "solution of (1)" means "nontrivial conjoined solution."

DEFINITION 2.4 A solution $Y = Y(x)$ of equation (1) is *oscillatory* if for each $a \in R^+$ there is a number b , $b \geq a$, such that $Y(b)$ is singular. The solution Y is *nonoscillatory* if it is not oscillatory. Equation (1) is oscillatory if it has at least one oscillatory solution; otherwise (1) is nonoscillatory.

In the finite dimensional case a solution Y of (1) is oscillatory if and only if $\det Y$ has an infinite number of zeros on R^+ . Of course, as noted above, $\det Y$ can have at most a finite number of zeros on any compact subset of R^+ . It is a consequence of Morse's generalization [30] of the Sturm separation theorem that if (1) has an oscillatory solution, then all solutions are oscillatory. The following example shows that this property does not carry over to the general B^*-algebra case.

EXAMPLE 2.4 It is easy to verify that if f is a continuous positive function on R^+ , then $\sin[\int_0^x f]$ and $\cos[\int_0^x f]$ are linearly independent solutions of the scalar equation $(y'/f)' + fy = 0$. Let r be a number such that $0 < r < 1$. Let $\{f_n\}$ be a sequence of positive continuous functions on R^+ such that for each positive integer n

$$\int_0^1 f_n(x)dx \geq r , \quad \int_0^n f_n(x)dx = \pi/2 ,$$

$$\text{and} \quad \int_0^\infty f_n(x)dx \leq 3\pi/4 .$$

Let $H = \ell_2$, and let P and Q be the infinite diagonal matrices defined by

$$P(x) = \text{diag}[1/f_1(x), 1/f_2(x), 1/f_3(x), \ldots] ,$$
$$Q(x) = \text{diag}[f_1(x), f_2(x), f_3(x), \ldots] .$$

The functions Z_1 and Z_2 given by

$$Z_1(x) = \text{diag}\{\sin[\int_0^x f_1], \sin[\int_0^x f_2], \sin[\int_0^x f_3], \ldots\}$$

$$Z_2(x) = \text{diag}\{\cos[\int_0^x f_1], \cos[\int_0^x f_2], \cos[\int_0^x f_3], \ldots\}$$

are nontrivial conjoined solutions of the equation $[P(x)Y']' + Q(x)Y = 0$. The solution Z_1 has no algebraic singularities on $(0,\infty)$, and is nonsingular on $[1,\infty)$. The solution Z_2 has an algebraic singularity at each integer $n \in R^+$. Thus Z_1 is nonoscillatory and Z_2 is oscillatory. Note that if the hypothesis $\int_0^1 f_n(x)dx \geq r$ for all positive integers n is deleted, then the sequence $\{f_n\}$ could have been chosen so that Z_1 is identically singular on R^+ with no singularity on $(0,\infty)$ being an algebraic singularity. This could be accomplished, for example, by having $\lim_{n\to\infty} \int_0^x f_n(t)dt = 0$ for each $x \in R^+$.

The next concept of oscillation which we want to consider is motivated by the vector-operator equation

$$(3) \qquad \ell[y] = [P(x)y']' + Q(x)y = 0$$

associated with equation (1), i.e., P and Q are the "coefficients" in (1) and $y : R^+ \to H$. Equation (3) is *oscillatory* if for each $a \in R^+$ there is a number b, $b \geq a$, and a solution y of (3), $y \not\equiv 0$, such that $y(0) = y(b) = 0$. Equation (3) is *nonoscillatory* if it is not oscillatory. The relationship between the oscillation of equations (1) and (3) in the finite dimensional case is easy to establish. In particular, if Y is the solution of (1) satisfying the initial conditions $Y(0) = 0$, $P(0)Y'(0) = I$, then $\det[Y(b)] = 0$, for some $b > 0$, if and only if there is a nontrivial solution y of (3) such that $y(0) = y(b) = 0$. Thus (3) is oscillatory if and only if (1) is oscillatory. The attempt to preserve this relationship in the general B^*-algebra case suggests the following more restricted concept of oscillation.

DEFINITION 2.5 A solution $Y = Y(x)$ of equation (1) is *algebraically oscillatory*, denoted A-*oscillatory*, if for each $a \in R^+$ there is a number b, $b \geq a$, such that $Y(b)$ has an algebraic singularity. The solution Y is *non*-A-*oscillatory* if it is not A-oscillatory. Equation (1) is A-oscillatory if it has at least one A-oscillatory solution; otherwise (1) is non-A-oscillatory.

By using the same proof as in the finite dimensional case, it can be shown that equation (3) is oscillatory if and only if the solution $Y = Y(x)$ of (1) satisfying $Y(0) = 0$, $P(0)Y'(0) = I$ is A-oscillatory. However, to see that the oscillation of (3) is not equivalent to the oscillation of (1) in the sense of Definition 2.4), or even to the A-oscillation of (1), in the general B^*-algebra case, consider again Example .4. The solution Z_1 has the initial values $Z_1(0) = 0$, $(0)Z_1'(0) = I$, and Z_1 is non-A-oscillatory. Thus the vector-operator equation associated with the given operator equation is nonoscillatory. On the other hand, the solution Z_2 is oscillatory, in fact, A-oscillatory. Thus the operator equation is -oscillatory.

It is obvious that if equation (1) is A-oscillatory, then it s oscillatory. It is an open question as to whether oscillation mplies A-oscillation. In this direction, however, we give an example of an equation having an oscillatory solution all of whose singularities are isolated and non-algebraic.

EXAMPLE 2.5 It is straightforward verification that for ny number a, $|a| > 1$, the function $y(x) = (a + \sin x)$ is solution of

$$y'' + (\sin x)(a + \sin x)^{-1}y = 0 .$$

et $H = \ell_2$, let $\{a_n\}$ be a decreasing sequence of positive umbers such that $\lim_{n\to\infty} a_n = 1$, and let Q be the infinite diagonal matrix given by

$$Q(x) = (\sin x)\mathrm{diag}[(a_1 + \sin x)^{-1},(a_2 + \sin x)^{-1},\ldots] .$$

he function $U(x) = \mathrm{diag}[(a_1 + \sin x),(a_2 + \sin x),\ldots]$ is a nontrivial conjoined solution of the equation

$$Y'' + Q(x)Y = 0 .$$

The solution U is nonsingular at each $x \in R^+$, except for $x = (4k + 3)\pi/2$, k a nonnegative integer. When

$x = (4k + 3)\pi/2$, $U = \text{diag}[(a_1 - 1),(a_2 - 1),\ldots]$, and while U is one-to-one, it does not have a bounded inverse since $(a_n - 1) \to 0$ as $n \to \infty$.

Hille [19, Chapter 9] considered equation (1) in the special case $P(x) \equiv I$ on R^+ . He has given a definition of oscillation which, stated in terms of our equation (1), is as follows.

DEFINITION 2.6 A solution $Y = Y(x)$ is *H-oscillatory* if Y^*PY' is not identically singular on R^+ and for each $a \in R^+$ there is a number b , $b \geq a$, such that $Y^*(b)P(b)Y'(b)$ has an algebraic singularity. The solution Y is *non-H-oscillatory* if it is not H-oscillatory. Equation (1) is H-oscillatory if it has at least one H-oscillatory solution; otherwise it is non-H-oscillatory.

If we consider equation (1) in the finite dimensional case $H = R_n$, and assume that $Q > 0$ on R^+ , then it can be shown that the solutions $Y = Y(x)$ of (1) satisfy the following separation property: between any two consecutive zeros of $\det Y$ there are at most n zeros of $\det PY'$, counting multiplicities, and between any two consecutive zeros of $\det PY'$ there are at most n zeros of $\det Y$, counting multiplicities. Thus we can conclude that Definitions 2.4, 2,5 and 2.6 are equivalent when $H = R_n$ and $Q > 0$. The following simple example illustrates that even in the finite dimensional case these definitions are not equivalent if the hypothesis $Q > 0$ is deleted.

EXAMPLE 2.6 Let $H = R_1 = R$, and let $q(x) = \sin(x) - \cos^2(x)$. The function $y(x) = e^{\sin x}$ is a solution of the scalar equation $y'' + q(x)y = 0$ on R^+ . Clearly y is a non-oscillatory solution of the equation and, by the Sturm eparation theorem, all solutions of the equation are nonoscillatory. Thus the equation is nonoscillatory in the sense of efinitions 2.4 and 2.5. On the other hand, since $y'(x) = (\cos x)e^{\sin x}$, it follows that yy' does oscillate, and so he equation is H-oscillatory.

An underlying objective of this paper is to extend the lassical theory of oscillation of the scalar equation (2) to the eneral operator equation (1), and therefore we shall want to void making any "sign" restrictions on the operator Q. Thus, s suggested by Example 2.6, Hille's definition of oscillation vill not be suitable for our purposes.

3. POSITIVE FUNCTIONALS

The methods and results of this paper involve the set of ositive linear functionals on the Banach algebra $\mathcal{B}$.

DEFINITION 3.1 A linear functional $g : \mathcal{B} \to \mathcal{R}$ is *positive* if $g(A^*A) \geq 0$ for all $A \in \mathcal{B}$. Equivalently, g is positive if $g(B) \geq 0$ whenever $B \in S$ and $B \geq 0$.

Let $\mathcal{G}$ denote the set of positive functionals on $\mathcal{B}$. C. E. Rickart [38] has shown that if $g \in \mathcal{G}$, then g is bounded (i.e., continuous), with $\|g\| = g(I)$, and satisfies a generalized Cauchy-Schwarz inequality

$$[g(A^*B)]^2 \leq g(A^*A)g(B^*B) \tag{4}$$

for all $A,B \in \mathcal{B}$. It follows from (4) that g is the zero functional if and only if $g(I) = 0$. If g is not the zero functional, then $g(I) > 0$, and, in general, $g(A) > 0$ whenever $A > 0$.

The set G is nonempty since it is obvious that the zero functional 0 is an element of G. It is easy to verify, however, that G contains elements in addition to the zero functional. For example, if $\alpha \in H$, $\alpha \neq 0$, then the functional g_α defined on B by

$$g_\alpha(A) = \langle A\alpha, \alpha \rangle \tag{5}$$

for all $A \in B$ is a positive functional with $\|g_\alpha\| = g(I) = \|\alpha\|^2 > 0$. Of course the zero functional is associated with the zero vector $0 \in H$. It can be shown that there are elements of G which are not "associated" with vectors in H through (5). For example, if $H = R_n$, then the functional "trace," denoted tr, and defined by $tr(A) = \sum_1^n a_{ii}$, is a positive functional which is not the associate of any vector $\alpha \in R_n$. In general, it can be verified that G is a positive cone in the space of continuous linear functionals on B.

Finally, since a positive functional g is continuous, it follows that

$$g\left[\int_a^x A(t)dt\right] = \int_a^x g[A(t)]dt , \quad a \geq 0 ,$$

whenever $A : R^+ \to B$ is integrable, and

$$g[B'(x)] = \{g[B(x)]\}'$$

whenever $B : R^+ \to B$ is differentiable.

4. OSCILLATION CRITERIA

In this section we develop oscillation criteria for equation (1). These criteria will involve the set G of positive functionals discussed in the last section. We shall also show how our criteria include a large number of well-known oscillation criteria as special cases, and so our approach can be viewed as a unification of the theory. We recall that the term "solution of (1)" shall be interpreted to mean "nontrivial, conjoined solution."

DEFINITION 4.1 A function $V : R^+ \to B$ is *L-admissible* if each of V and PV' is continuously differentiable on R^+ and

$$V^*[PV'] \equiv [PV']^*V .$$

Our first result is a "Picone type" identity for equation (1). For a complete discussion of the Picone and related identities, the reader is referred to Reid [36, Chapter VII] and to Swanson [41,42].

THEOREM 4.1 Let $g \in G$, and let $f : R^+ \to R$ be piecewise continuously differentiable. If $V : R^+ \to B$ is an L-admissible function which is nonsingular on an interval $J \subseteq R^+$, then

$$g\{(f'I - fV'V^{-1})^*P(f'I - fV'V^{-1})\} + \{f^2g[PV'V^{-1}]\}'$$
$$= f'^2g[P] - f^2g[Q] + f^2g\{L[V]V^{-1}\} \quad \text{on } J .$$

The proof of this identity can be established by a straightforward verification. We use this version of Picone's identity to obtain the following oscillation criterion for equation (1).

THEOREM 4.2 If for each $a \in R^+$ there is a number b, $b > a$, an element $g \in G$, $g \neq 0$, and a piecewise continuously differentiable function $f : R^+ \to R$ such that $f(a) = f(b) = 0$, $f \not\equiv 0$ on $[a,b]$, and

$$(6) \qquad \int_a^b \{f'^2(x)g[P(x)] - f^2(x)g[Q(x)]\}dx \le 0 ,$$

then all solutions of equation (1) are oscillatory.

PROOF Suppose there is a solution $V = V(x)$ of (1) such that V is nonoscillatory. Then there is a number $a \in R^+$ such that V is nonsingular on $[a,\infty)$.

It is well-known that the scalar equation (2) has a solution with at least two zeros on an interval $[a,b]$ if and only if there is a piecewise continuously differentiable function $f : R^+ \to R$ such that $f(a) = f(b) = 0$, $f \not\equiv 0$ on $[a,b]$, and

$$\int_a^b [f'^2(x)p(x) - f^2(x)q(x)]dx \le 0 .$$

From this fact we can conclude that there are numbers c and d, $a \le c < d \le b$, and a nontrivial solution $u = u(x)$ of the scalar equation

$$(g[P(x)]u')' + g[Q(x)]u = 0$$

on $[a,b]$ such that $u(c) = u(d) = 0$. From Theorem 4.1, we have that

$$\int_c^d g\{(u'I - uV'V^{-1})^*P(u'I - uV'V^{-1})\}dx$$

$$= \int_c^d \{u'^2g[P] - u^2g[Q]\}dx$$

$$= -\int_c^d \{(g[P]u')' + g[Q]u\}udx$$

$$= 0 .$$

Now, since u is nontrivial and $u(c) = 0$, we must have $u'(c) \neq 0$ which implies that there is a nondegenerate interval $[c,c') \subset [c,d]$ such that $u'I - uV'V^{-1}$ is nonsingular on $[c,c')$. Therefore, $(u'I - uV'V^{-1})^*P(u'I - uV'V^{-1}) > 0$ on $[c,c')$ since $P > 0$, and this leads to the contradiction

$$0 < \int_c^{c'} g\{(u'I - uV'V^{-1})^*P(u'I - uV'V^{-1})\}dx$$

$$\leq \int_c^d g\{(u'I - uV'V^{-1})^*P(u'I - uV'V^{-1})\}dx = 0 .$$

The proof of Theorem 4.2 suggests the following oscillation criterion for equation (1).

THEOREM 4.3 If there is a $g \in G$ such that the scalar equation

$$(7) \qquad (g[P(x)]y')' + g[Q(x)]y = 0$$

is oscillatory, then all solutions of equation (1) are oscillatory.

PROOF Simply use an oscillatory solution u of (7) to construct a piecewise continuously differentiable function f such that it, together with the given functional $g \in G$, satisfies the hypotheses of Theorem 4.2.

As a consequence of Theorem 4.3, we can consider the question of the oscillation of equation (1) in terms of the oscillation of an associated scalar equation of the form (2). Thus *any* of the very large number of well-known oscillation criteria for (2) can be used to determine a corresponding oscillation criterion for (1). The following corollary is a generalization of the Leighton-Wintner oscillation criterion, and it is a simple example of the type of criteria which can be obtained for (1) through Theorem 4.3.

COROLLARY If there is a $g \in G$ such that

$$\int_0^\infty \frac{dx}{g[P(x)]} = \int_0^\infty g[Q(x)]dx = \infty \,, \tag{8}$$

then all solutions of equation (1) are oscillatory.

Included in this Corollary are most of the well-known oscillation criteria for (1) in both the finite and infinite dimensional cases. We demonstrate this statement by giving some specific examples.

EXAMPLE 4.1 (Hayden and Howard [17, Theorem 2] and Howard [21, Theorem 1].) In equation (1) let $P(x) \equiv I$, and let $K(x) = \int_0^x Q(t)dt$. If

$$\inf_{\|\alpha\|=1} \langle K(x)\alpha,\alpha\rangle \to \infty \quad \text{as} \quad x \to \infty \,,$$

then all solutions are oscillatory.

PROOF Fix *any* $\beta \in H$, $\|\beta\| = 1$, and let g_β be the positive functional associated with β using (5). Then $g_\beta(P) = g_\beta(I) = \|\beta\|^2 = 1$, and

$$\inf_{\|\alpha\|=1} \langle K(x)\alpha,\alpha\rangle \le g_\beta[K(x)] = \int_0^x g_\beta[Q(t)]dt .$$

Thus the given hypotheses imply $\int_0^\infty \frac{1}{g_\beta(P)} = \int_0^\infty g_\beta(Q) = \infty$, and the result follows from the Corollary.

EXAMPLE 4.2 (Etgen [10, Theorem 2].) Let $H = R_n$, and let $P(x) \equiv I$ in equation (1). If $\int_0^\infty tr[Q(x)]dx = \infty$, then all solutions are oscillatory.

PROOF The functional "trace" is a positive functional, and $tr[P(x)] = tr(I) = n$. Thus the given hypotheses imply that $\int_0^\infty \frac{1}{tr(P)} = \int_0^\infty tr(Q) = \infty$, and this result is a special case of the Corollary.

EXAMPLE 4.3 (Noussair and Swanson [32, Theorem 2].) Let $H = R_n$. If there exists an integer i , $1 \le i \le n$, such that the diagonal elements p_{ii} and q_{ii} of P and Q , respectively, satisfy

$$\int_0^\infty \frac{dx}{p_{ii}(x)} = \int_0^\infty q_{ii}(x)dx = \infty ,$$

then all solutions of equation (1) are oscillatory.

PROOF Let ε_i be the vector with a 1 in the i-th position and zeros elsewhere. Then the given hypotheses imply that $\int_0^\infty \frac{1}{g_{\varepsilon_i}(P)} = \int_0^\infty g_{\varepsilon_i}(Q) = \infty$, so that the Corollary applies.

EXAMPLE 4.4 (W. Allegretto and L. Erbe [3, Corollary 1].) Let $H = R_n$. Let $S_{k,n}$ denote the collection of strictly increasing sequences of k integers chosen from the set {1,2, ...,n} . For any $n \times n$ matrix A , and any $\sigma(k) = \{i_1, i_2, \ldots, i_k\} \in S_{k,n}$, let $\sum_\sigma A$ denote the sum of the entries of the $k \times k$ submatrix of A obtained by deleting all rows and columns of A except the $i_1, i_2, \ldots, i_k$ rows and columns. If there exists $\sigma(k) \in S_{k,n}$ such that

$$\int_0^\infty [\sum_\sigma P(x)]^{-1} dx = \int_0^\infty [\sum_\sigma Q(x)] dx = \infty ,$$

then all solutions of equation (1) are oscillatory.

PROOF Let α be the vector with "ones" in the $i_1, i_2, \ldots,$ i_k positions and zeros elsewhere, and let g_α be the positive functional associated with α using (5). Then $g_\alpha(A) = \sum_\sigma A$ for all $n \times n$ matrices A . Thus the hypotheses can be restated as $\int_0^\infty \frac{1}{g_\alpha(P)} = \int_0^\infty g_\alpha(A) = \infty$ and the Corollary applies.

In a similar manner, the oscillation criteria obtained by such authors as Kartsatos [22], Kreith [23], Swanson [40] and Tomastik [44] can be demonstrated to be special cases of the Corollary by making suitable choices for the positive functional g .

As suggested by the above examples, most of the oscillation criteria for equation (1) in the finite and infinite dimensional cases are generalizations of the Leighton-Wintner oscillation criterion for the scalar equation (2), i.e., most of the oscillation criteria involve assumptions of the form (8). In contrast, Theorem 4.3 can also be used to obtain oscillation criteria for equation (1) of the Hille-Wintner type where it is assumed that " $\int_0^\infty Q(x)dx$ is convergent." As a simple example of this type of criteria, we have:

EXAMPLE 4.5 Let $P(x) \equiv I$ on R^+. If there exists a $g \in G$ such that $\int_0^\infty g[Q(x)]dx$ converges (possibly just conditionally), and if

$$\liminf_{x\to\infty} x \int_x^\infty g[Q(t)]dt > \frac{1}{4},$$

then all solutions of equation (1) are oscillatory.

The next result is an extension of Theorem 4.2 which has been established by the authors [11]. The motivation for this result and the method of proof are contained in the work of D. B. Hinton [20].

THEOREM 4.4 If there is a positive continuous function h on R^+ such that for each $a \in R^+$

(i) $\int_a^\infty h(x)dx = \infty$,

(ii) $$\lim_{k\to\infty} \frac{\int_a^{t_k} \{h^2(x)g_k[P(x)] - [\int_x^{t_k} h(s)ds]^2 g_k[Q(x)]\}dx}{[\int_a^{t_k} h(s)ds]^2} = -\infty$$

for some sequence $\{t_k\}$ in R^+ with $\lim_{k\to\infty} t_k = \infty$, and some sequence of positive functionals $\{g_k\}$ in G with the property that there exists a positive number M such that $\|g_k\| \le M$ for all positive integers k, then all solutions of equation (1) are oscillatory.

Theorem 4.4 leads to a variety of oscillation criteria for equation (1) which are more general than those indicated in

Examples 4.1 - 4.5 in the sense that they cannot be obtained as Corollaries of Theorem 4.3. The following two corollaries are examples of the type of oscillation criteria which result from Theorem 4.4.

COROLLARY 1 If there is a positive continuous function h on R^+ such that for each $a \in R^+$

(i) $\int_a^\infty h(x)dx = \infty$,

(ii)

$$\lim_{k\to\infty} \frac{\int_a^{t_k} \{h^2(x)\langle P(x)\xi_k,\xi_k\rangle - [\int_x^{t_k} h(s)ds]^2\langle Q(x)\xi_k,\xi_k\rangle\}dx}{[\int_a^{t_k} h(s)ds]^2} = -\infty$$

for some sequence $\{t_k\}$ in R^+ with $\lim_{k\to\infty} t_k = \infty$, and some sequence $\{\xi_k\}$ of unit vectors in H , then every solution of equation (1) is oscillatory.

COROLLARY 2 If $\int_0^\infty \frac{dx}{|||P(x)|||} = \infty$, and if for each $a \in R^+$

(i)

$$\limsup_{t\to\infty} \left\{ \sup_{\|\xi\|=1} \frac{\langle [\int_a^t \{\int_x^t \frac{ds}{|||P(s)|||}\}^2 Q(x)dx]\xi,\xi\rangle}{[\int_a^t \frac{ds}{|||P(s)|||}]^2} = \infty \right\} ,$$

then all solutions of equation (1) are oscillatory. In the special case $H = R_n$, condition (i) can be expressed as

$$\limsup_{t\to\infty} \frac{\nu[\int_a^t \{\int_x^t \frac{ds}{|||P(s)|||}\}^2 Q(x)dx]}{[\int_a^t \frac{ds}{|||P(s)|||}]^2} = \infty ,$$

where $\nu[A]$, A an $n \times n$ symmetric matrix, denotes the maximum eigenvalue of A .

5. EXTENSIONS

The methods and results of the preceding section can be extended in a variety of ways. We conclude this paper by briefly indicating several areas where the methods can be applied.

The first extension involves the coefficient function $P : \mathcal{R}^+ \to S$ in equation (1). Up to this point we have been assuming that $P(x) > 0$ for all $x \in \mathcal{R}^+$, but with only minor modifications of the results in Section 4, this requirement can be relaxed to $P(x) \geq 0$ for all $x \in \mathcal{R}^+$. As an example of the type of modification which is required by relaxing the positive definiteness of P, the quadratic inequality (6) will have to be strengthened to a strict inequality. A second extension involves a nonlinear analog of the linear differential operator L. Nonlinear matrix equations, as well as linear and nonlinear matrix differential inequalities have been investigated by a number of authors. See, for example, Allegretto and Erbe [3], Etgen [8,10], Kartsatos [22], Kreith [23], Noussair and Swanson [32], Swanson [40], and Tomastik [44]. An examination of these papers shows that the nonlinear operators are defined in such a manner that the methods developed for linear equations of the form (1) can be applied. In the discussion which follows we combine these two extensions into one development.

Let $P,Q : \mathcal{R}^+ \times \mathcal{B} \times \mathcal{B} \to S$ be continuous with $P(x,A,B) \geq 0$ for all $(x,A,B) \in \mathcal{R}^+ \times \mathcal{B} \times \mathcal{B}$. Let Γ denote the collection of functions $Y : \mathcal{R}^+ \to \mathcal{B}$ such that Y and $P(x,Y,Y')Y'$ are continuously differentiable and

$$Y^*[P(x,Y,Y')Y'] \equiv [P(x,Y,Y')Y']^*Y$$

on $\mathcal{R}^+$. Let $\mathcal{L}$ be the nonlinear differential operator defined on Γ by

$$L[Y] = [P(x,Y,Y')Y']' + Q(x,Y,Y')Y$$

and consider the differential inequality

$$(9) \qquad Y^* L[Y] \le 0 \ .$$

As an analog of the work in the preceding section, we are concerned with the oscillation of solutions of (9). The concepts of oscillation, nonoscillation, etc., of solutions of (9), as well as the oscillation or nonoscillation of (9) itself, remain as defined in Section 2. Since the continuation problem is not under consideration, we assume that all solutions of (9) can be continued over R^+.

Just as in the linear case, a straightforward verification establishes the following nonlinear version of the Picone type identity of Theorem 4.1.

THEOREM 5.1 Let $g \in G$, and let $f : R^+ \to R$ be piecewise continuously differentiable. If $V \in \Gamma$ is nonsingular on an interval $J \subseteq R^+$, then

$$\begin{aligned} g\{[f'I - fV'V^{-1}]^* P(x,V,V')[f'I - fV'V^{-1}]\} \\ + \{f^2 g[P(x,V,V')V'V^{-1}]\}' \\ = f'^2 g[P(x,V,V')] - f^2 g[Q(x,V,V')] + f^2 g\{L[V]V^{-1}\} \end{aligned}$$

on J.

The next result corresponds to Theorem 4.2. As indicated above, the relaxation of the positive definiteness of P on R^+ requires a strengthening of the inequality corresponding to (6), but, in so doing, the proof is actually simplified.

THEOREM 5.2 Every solution of (9) is oscillatory if for each $a \in R^+$ there is a number b, $b > a$, an element $g \in G$, and a piecewise continuously differentiable function $f : R^+ \to R$ such that $f(a) = f(b) = 0$, $f \not\equiv 0$ on $[a,b]$, and

$$\int_a^b \{f'^2 g[P(x,V,V')] - f^2 g[Q(x,V,V')]\}dx < 0$$

for every $V \in \Gamma$ such that V is nonsingular on $[a,\infty)$.

Now, by using the approach suggested in Section 4, nonlinear versions of Theorem 4.3 and its Corollary are easy to establish. Then, as in the linear case, these results lead to oscillation criteria for (9) which include the criteria contained in [3], [8], [10], [22], [23], [32], [40] and [44] as special cases. Finally, it can be verified that the analog of the more general result contained in Theorem 4.4 also holds for (9).

The next extension is concerned with higher order linear equations. Oscillation criteria for even order selfadjoint scalar differential equations of the form

$$(10) \qquad [p(x)y^{(n)}]^{(n)} + (-1)^{n-1}q(x)y = 0$$

on R^+, where $n \geq 1$, $p,q : R^+ \to R$ are continuous, and $p(x) > 0$ for all $x \in R^+$, have been developed by D. B. Hinton [20]. As remarked previously, Theorem 4.4 is motivated by Hinton's work. The vector-operator analog of (10) is

$$(11) \qquad [P(x)y^{(n)}]^{(n)} + (-1)^{n-1}Q(x)y = 0,$$

where $P,Q : R^+ \to S$ are continuous and $P(x) > 0$ for all $x \in R^+$. Equation (11) is oscillatory if for each $a \in R^+$ there is a number b, $b > a$, and a solution $y : R^+ \to H$ such that

$$y(a) = y'(a) = \cdots = y^{(n-1)}(a) = 0$$
$$= y(b) = y'(b) = \cdots = y^{(n-1)}(b) .$$

The following theorem is an extension of Corollary 1, Theorem 4.4, and it generalizes Hinton's theorem [20, Theorem 3].

THEOREM 5.3 If there is a positive continuous function h on R^+ such that for each $a \in R^+$

(i) $\int_a^\infty x^{n-1}h(x)dx = \infty$,

(ii)

$$\lim_{k\to\infty} \left\{ \left(\int_a^{t_k} \{ \langle P(x)\xi_k,\xi_k \rangle [(n-1)!h(x)dx]^2 - \langle Q(x)\xi_k,\xi_k \rangle [\int_a^{t_k} (s-x)^{n-1}h(s)ds]^2 \}dx \right) \middle/ \left([\int_a^{t_k} s^{n-1}h(s)ds]^2 \right) \right\} = -\infty$$

for some sequence $\{t_k\}$ in R^+ with $\lim_{k\to\infty} t_k = \infty$, and some sequence $\{\xi_k\}$ of unit vectors in H , then equation (11) is oscillatory.

The final extension concerns elliptic partial differential equations of the form

$$(12) \quad \ell[U] = \sum_{i,j=1}^{n} D_i[P_{ij}(x)D_jU] + Q(x)U = 0 \ , \ (D_i = \frac{\partial}{\partial x_i}),$$

on an unbounded domain G in n-dimensional Euclidean space R_n , where P_{ij} , $i,j = 1,2,\ldots,n$, and Q are continuous, symmetric, $m \times m$, matrix-valued functions on G , and the $mn \times mn$ matrix-valued function $(P_{ij}(x))$ is nonnegative definite on G . Oscillation criteria for (12), as well as for corresponding nonlinear versions of (12) and for differential inequalities of the form $U^{*}\ell[U] \le 0$, have been developed by Allegretto and Swanson [4], Kreith and Travis [25], Noussair [34], Headly and Swanson [18], and Swanson [39], as well as a number of other authors. The methods employed by these authors are related to corresponding methods and results for ordinary differential equations of the form (1). It is clear, therefore, that the techniques developed in this paper can also be applied to (12), thereby extending and unifying many of the oscillation criteria for elliptic partial differential equations. The following theorem is the analog of Theorem 4.2, and it is an example of the type of oscillation criteria which can be obtained for (12) using the methods of this paper.

THEOREM 5.4 Equation (12) is oscillatory in the unbounded domain G if:

(i) for arbitrary $a > 0$, the set $G_a = G \cap \{x \in R_n : \|x\| > a\}$ contains a nonempty regular bounded domain M_a, and

(ii) there exists a $g \in G$ (the set of positive functionals on the $m \times m$ matrices) and a piecewise continuously differentiable function f on the closure of M_a such that

$f \equiv 0$ on ∂M_a and

$$\int_{M_a} \Big[\sum_{i,j=1}^{n} g(P_{ij})(D_i f)(D_j f) - f^2 g(B) \Big] < 0 .$$

REFERENCES

1. Ahlbrandt, C. D. Disconjugacy criteria for self-adjoint differential systems, *J. Diff. Equa.* 6(1969), pp. 271-295.

2. Ahmad, S. and A. C. Lazer. Component properties of second order linear systems, *Bull. Amer. Math. Soc.* 82(1976), pp. 287-289.

3. Allegretto, W. and L. Erbe. Oscillation criteria for matrix differential inequalities, *Canad. Math. Bull.* *16*(1973), pp. 5-10.

4. Allegretto, W. and C. A. Swanson. Oscillation criteria for elliptic systems, *Proc. Amer. Math. Soc.* *27*(1971), pp. 325-330.

5. Atkinson, F. V. "Discrete and Continuous Boundary Problems." Academic Press, New York, 1964.

6. Barrett, J. H. Oscillation theory of ordinary linear differential equations, *Adv. Math.* *3*(1969), Academic Press, New York.

7. Coppel, W. A. "Disconjugacy." Lecture Notes in Mathematics No. 220, Springer-Verlag, Berlin-Heidelberg-New York, 1971.

8. Etgen, G. J. Oscillatory properties of certain nonlinear matrix differential systems, *Trans. Amer. Math. Soc.* *122* (1966), pp. 289-310.

9. Etgen, G. J. Two point boundary problems for second order matrix differential systems, *Trans. Amer. Math. Soc.* *149* (1970), pp. 119-132.

10. Etgen, G. J. Oscillation criteria for nonlinear second order matrix differential equations, *Proc. Amer. Math. Soc.* *27* (1971), pp. 259-267.

11. Etgen, G. J. and R. T. Lewis. The oscillation of ordinary differential equations in a B*-algebra, submitted for publication.

12. Etgen, G. J. and J. F. Pawlowski. Oscillation criteria for second order selfadjoint differential systems, *Pacific J. Math.* *66*(1976), pp. 99-110.

13. Etgen, G. J. and J. F. Pawlowski. A comparison theorem and oscillation criteria for second order differential systems, *Pacific J. Math.*, to appear.

14. Glazman, I. M. "Direct Methods of Qualitative Spectral Analysis of Singular Differential Operators." Israel Program for Scientific Translation, Jerusalem, 1965.

15. Hartman, P. Selfadjoint, nonoscillatory systems of ordinary second order linear differential equations, *Duke Math. J.* *24*(1957), pp. 25-35.

16. Hartman, P. "Ordinary Differential Equations." John Wiley and Sons, New York, 1964.

17. Hayden, T. L. and H. C. Howard. Oscillation of differential equations in Banach spaces, *Annali di Mathematica Pura ed Applicata* *85*(1970), pp. 383-394.

18. Headly, V. B. and C. A. Swanson. Oscillation criteria for elliptic equations, *Pacific J. Math.* *27*(1968), pp. 501-506.

19. Hille. E. "Lectures on Ordinary Differential Equations." Addison-Wesley Publishing Company, Reading, Mass., 1969.

20. Hinton, D. B. A criterion for n-n oscillations in differential equations of order 2n , *Proc. Amer. Math. Soc.* *19* (1968), pp. 511-518.

21. Howard, H. C. Oscillation criteria for matrix differential equations, *Canad. J. Math.* *19*(1967), pp. 184-199.

22. Kartsatos, A. G. Oscillation of nonlinear systems of matrix differential equations, *Proc. Amer. Math. Soc.* *30*(1971), pp. 97-101.

23. Kreith, K. Oscillation criteria for nonlinear matrix differential equations, *Proc. Amer. Math. Soc.* *26*(1970), pp. 270-272.

24. Kreith, K. "Oscillation Theory." Lecture Notes in Mathematics No. 324, Springer-Verlag, Berlin-Heidelberg-New York, 1973.

25. Kreith, K. and C. C. Travis. Oscillation criteria for selfadjoint elliptic equations, *Pacific J. Math.* *41*(1972), pp. 743-753.

26. Lewis, R. T. Oscillation and nonoscillation criteria for some selfadjoint even order linear differential operators, *Pacific J. Math.* *51*(1974), pp. 221-234.

27. Lewis, R. T. The existence of conjugate points for selfadjoint differential equations of even order, *Proc. Amer. Math. Soc.* *56*(1976), pp. 162-166.

28. Lewis, R. T. Conjugate points of vector-matrix differential equations, *Trans. Amer. Math. Soc.*, to appear.

29. Martynov, V. V. Conditions for discreteness and continuity of the spectrum in the case of a selfadjoint system of differential equations of even order, *Dif. Uravneniya 1*(1965), pp. 1578-1591.

30. Morse, M. A generalization of the Sturm separation and comparison theorems in n-space, *Math. Ann. 103*(1930), pp. 52-69.

31. Morse, M. "The Calculus of Variations in the Large." *Amer. Math. Soc. Colloq. Publ. 18*, Amer. Math. Soc., Providence, Rhode Island, 1934.

32. Noussair, E. S. and C. A. Swanson. Oscillation criteria for differential systems, *J. Math. Anal. and Appl. 36*(1971), pp. 575-580.

33. Noussair, E. S. Differential equations in Banach spaces, *Bull. Australian Math. Soc. 9*(1973), pp. 219-226.

34. Noussair, E. S. Oscillation theory of elliptic equations of order 2m , *J. Diff. Equa. 10*(1971), pp. 100-111.

35. Reid, W. T. Riccati matrix differential equations and non-oscillation criteria for associated linear differential systems, *Pacific J. Math. 13*(1963), pp. 665-685.

36. Reid, W. T. "Ordinary Differential Equations." John Wiley and Sons, New York, 1971.

37. Reid, W. T. "Riccati Differential Equations." Mathematics in Science and Engineering, Volume 86, Academic Press, New York, 1972.

38. Rickart, C. E. "General Theory of Banach Algebras." Van Nostrand, New York, 1960.

39. Swanson, C. A. "Comparison and Oscillation Theory of Linear Differential Equations." Academic Press, New York, 1968.

40. Swanson, C. A. Oscillation criteria for nonlinear matrix differential inequalities, *Proc. Amer. Math. Soc. 24*(1970), pp. 824-827.

1. Swanson, C. A. Remarks on Picone's identity and related identities, *Atti. Accad. Naz. Lincer 11*(1972), pp. 1-15.

2. Swanson, C. A. Picone's identity, *Rendiconti di Matematica 8 Serie VI*(1975), pp. 373-397.

3. Tomastik, E. C. Singular quadratic functionals of n dependent variables, *Trans. Amer. Math. Soc. 124*(1966), pp. 60-76.

4. Tomastik, E. C. Oscillation of nonlinear matrix differential equations of second order, *Proc. Amer. Math. Soc. 19*(1968), pp. 1427-1431.

5. Tomastik, E. C. Oscillation of systems of second order differential equations, *J. Diff. Equa. 9*(1971), pp. 436-442.

6. Tomastik, E. C. Principal quadratic functionals, *Trans. Amer. Math. Soc. 218*(1976), pp. 297-309.

7. Williams, C. M. "Oscillation Phenomena for Linear Differential Systems in a B*-Algebra," Ph.D. Dissertation, The University of Oklahoma, 1971.

8. Yakubovic, V. A. Oscillation properties of solutions of linear canonical systems of differential equations, *Doklady Akad. Nauk SSR 124*(1959), pp. 533-536.

9. Yakubovic, V. A. Oscillation and non-oscillation conditions for canonical linear sets of simultaneous differential equations, ibid, pp. 994-997.

0. Yakubovic, V. A. Oscillation properties of solutions of canonical equations, *Mat. Sb. 56*(1962), pp. 3-42.

MATRIX EQUATIONS IN THE SIEGEL DISK

Raymond Redheffer

University of California at Los Angeles
Los Angeles, California

1 THE SIEGEL DISK

The sets of nonnegative reals, complex n-vectors, and complex n by n matrices are denoted respectively by R^+, C^n and M^n. As norms for $\xi \in C^n$ and $z \in M^n$, respectively, we use

$$(1) \qquad |\xi| = |\xi^*\xi|^{\frac{1}{2}}, \quad \|z\| = \sup_{|\xi|=1} |z\xi| .$$

Hence, the equation $\|z\| \le 1$ is equivalent to

$$(2) \qquad \sup(\xi^* z^* z\xi) \le 1 \qquad (|\xi| = 1)$$

and this in turn is equivalent to $z^* z \le I$, in the usual ordering, where I is the n by n identity matrix.

ISBN 0-12-632250-3

The set of matrices $z \in M_n$ satisfying $z^*z \leq I$ is called the *Siegel disk*. A point z is on the boundary of the Siegel disk if equality holds for some ξ in (2). Hence, the boundary is the set of points z such that $\|z\| \leq 1$ and

$$\det(z^*z - I) = 0 .$$

The set of unitary matrices $z \in M_n$ is the *Šilov boundary*. This is a subset of the boundary in which broad classes of functions $f(z)$ for $\|z\| \leq 1$ can be counted upon to assume their maximum.

Some interesting theorems on matrix differential equations, due to Reid and others, can be interpreted as stating that the Siegel disk is invariant. That is, if $\|u(0)\| \leq 1$, then also $\|u(t)\| \leq 1$ for $t > 0$. Other theorems assert invariance of the boundary or the Šilov boundary. When the underlying differential equation is specialized to be a Riccati equation, these invariance theorems have an important bearing on the theory of multiple transmission lines, scattering matrices, and radiative transport.

2 SET-VALUED FUNCTIONS

Let u be a function from $\mathbb{R}^+$ to M^n. A right-hand derived value is an element $q \in M^n$ such that

$$\lim_{n\to\infty} \frac{u(t + h_n) - u(t)}{h_n} = q$$

for some sequence $\{h_n\}$, $h_n > 0$, $\lim h_n = 0$. The set of all such values is denoted by $u'(t)$; clearly $u'(t)$ is closed. Thus, u' is a function from $\mathbb{R}^+$ to the set of closed subsets of M^n.

A proposition is said to hold mod E if it holds except in a set of cardinality less than the continuum. No harm is done if the reader interprets mod E to mean "except in an enumerable set" and the letter E was chosen with this interpretation in mind.

DEFINITION 1 The class D is the class of continuous functions $u : R^+ \to M^n$ which satisfy one of the following conditions:

(i) u is absolutely continuous, or

(ii) $u'(t)$ is nonempty mod E .

To formulate a differential equation $u' = f(t,u)$ in this context, let us agree that $f(t,q)$ is a subset of M^n for each $t \in R^+$ and $q \in M^n$. Then the hypothesis

$$(3) \qquad u'(t) \cap f[t,u(t)] \neq \emptyset$$

reduces to $u'(t) = f[t,u(t)]$ whenever both sets are singletons. Furthermore, this hypothesis is equivalent to $u'(t) \in f[t,u(t)]$ whenever $u'(t)$ exists in the ordinary sense and f is multi-valued. And finally, the hypothesis reudces to $f[t,u(t)] \in u'(t)$ when u' is a set and f is single-valued. This is the matrix analog of an equation involving Dini derivatives for functions $\mathbb{R} \to \mathbb{R}$.

3 AN ELEMENTARY INVARIANCE THEOREM

The quantity [z] defined for $z \in M^n$ by

$$(4) \qquad [z] = \sup \operatorname{Re} \xi^* z \xi \qquad (|\xi| = 1)$$

satisfies $[\lambda z] = \lambda[z]$ for $\lambda \geq 0$ and $[p + q] \leq [p] + [q]$;

hence, it is a Kamke norm. In the following theorem, as elsewhere in this paper, ε denotes a positive constant which can be arbitrarily small.

THEOREM 1 With $u \in D$ and f as above suppose the two conditions

$$1 + \varepsilon > \|p\| > 1 , \qquad q \in f(t,p)$$

together imply $[p^* q] \leq 0$. Furthermore, suppose almost all the sets

$$u'(t) \cap f[t,u(t)]$$

are nonempty. Then the Siegel disk is invariant.

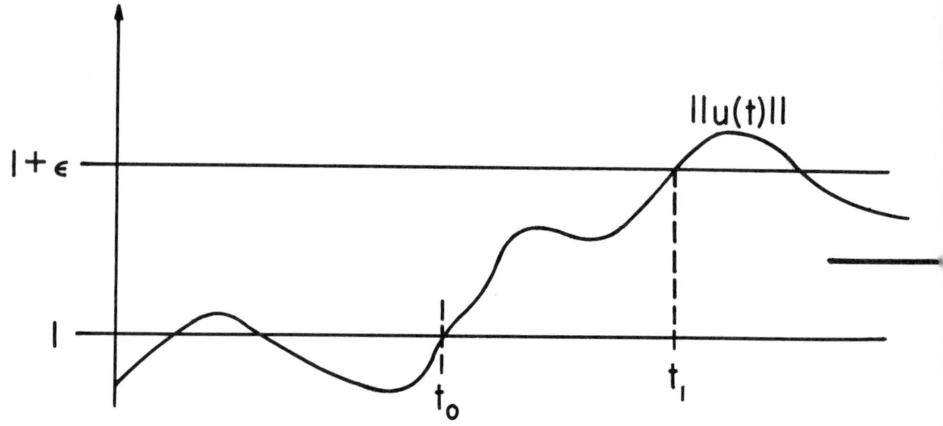

Figure 1

The proof is simple. If $\|u(t)\| \geq 1 + \varepsilon$ for some $t > 0$ we could choose first t_1 and then $t_0 < t_1$ as suggested by Figure 1. Fix $\xi \in C^n$ so that $|\xi| = 1$ and $|u(t_1)\xi| > 1$ and define

$$v(t) = |u(t)\xi|^2 = \xi^* u^*(t)u(t)\xi \quad .$$

At any point $t \in (t_0, t_1)$ where $p = u(t)$ and $q \in u'(t)$, the lower right Dini derivative satisfies

$$D_+ v(t) \le \xi^*(p^*q + q^*p)\xi \le [p^*q] \ .$$

Under either hypothesis (i) or (ii) of the class D it follows that v is weakly decreasing, hence

$$v(t_1) \le v(t_0) \le \|u(t_0)\| = 1 \ ,$$

and this contradicts the choice of ξ.

4 THE TANGENT CONDITION

Let p be a point on the boundary of the Siegel disk, so that $\|p\| = 1$. It is said that f satisfies the *tangent condition* at p if for some $q \in f(t,p)$ we have

$$(5) \qquad \liminf_{h\to 0+} \frac{\|p + hq\| - 1}{h} \le 0$$

(see Figure 2). We use lim inf in (5) because this agrees with the well-known tangent condition of Nagumo. Actually the lim inf exists as a limit, and this fact is used below.

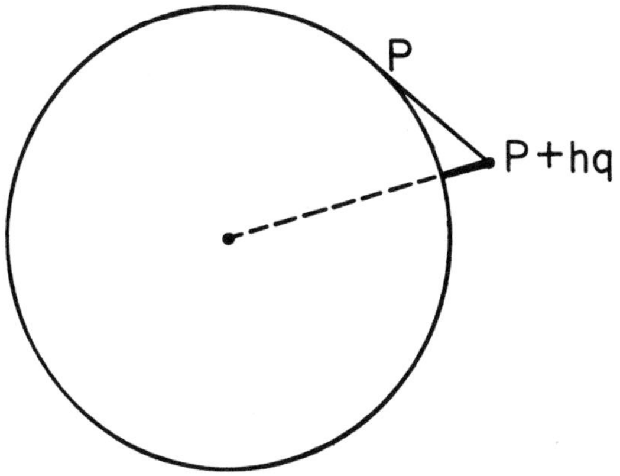

Figure 2

The expression following the lim inf in (5) is

$$\tfrac{1}{2} \sup_{|\xi|=1} \xi^* \left(p^* q + q^* p - \frac{I - p^* p}{h} \right) \xi \tag{6}$$

aside from a term $O(h)$. Since the term involving h makes a nonpositive contribution, the tangent condition certainly holds if $[p^* q] \le 0$. This gives a geometric interpretation of the main hypothesis in Theorem 1. However, a much sharper result is obtained if the term $I - p^* p$ in (6) is not just discarded, but is exploited. In this sharper formulation, the tangent condition (5) is found to be equivalent to $[p,q] \le 0$ where $[p,q]$ is defined as follows:

DEFINITION 2 Let $x \in M^n$, $y \in M^n$ and let $\sigma(x)$ denote the set of unit vectors $\xi \in C^n$ for which $x^* x\xi = \|x^* x\|\xi$. Then

$$[x,y] = \sup_{\xi \in \sigma(x)} \frac{\mathrm{Re}\,\xi^* x^* y \xi}{\|x\|}, \qquad (x \ne 0).$$

5 A SHARPER THEOREM

Without any attempt at maximum generality, we introduce a class U of "uniqueness functions" as follows:

DEFINITION 3 The above function f belongs to the class U if there exist positive constants μ and ε such that the conditions

$$1 + \varepsilon > \|x\| \ge 1, \quad 1 + \varepsilon > \|y\| \ge 1, \quad \tilde{x} \in f(t,x), \quad \tilde{y} \in f(t,y)$$

together imply $[x - y, \tilde{x} - \tilde{y}] \le \mu\|x - y\|$.

The next result, which we state without proof, is a modification of [1], Theorem 1. See also [2]:

THEOREM 2 Let $u \in D$ and $f \in U$. Whenever $\|p\| = 1$ suppose there exists $q \in f(t,p)$ such that $[p,q] \le 0$. Suppose, further, that almost all the sets

$$u'(t) \cap f[t,u(t)]$$

are nonempty. Then the Siegel disk is invariant.

Aside from the hypothesis $f \in U$, Theorem 2 is much sharper than Theorem 1. The main difference is that the condition $[p^* q] \le 0$ of Theorem 1 is replaced by the weaker condition $[p,q] \le 0$. Moreover, this hypothesis was required in Theorem 1 for all $q \in f(t,p)$, while the present weaker hypothesis is required for just one $q \in f(t,p)$. A different element $\tilde{q} \in u'(t)$ can be used to satisfy the differential equation $\tilde{q} \in f[t,u(t)]$ even if $u(t) = p$. When asked why he did not use quaternions, Maxwell replied, "Can one plow with an ox and an ass together?" In the theory of differential equations with set-valued functions, one can!

6 THE RICCATI EQUATION

We apply Theorem 2 to the relation

$$(7) \qquad a(t) + b(t)u(t) + u(t)d(t) + u(t)c(t)u(t) \in u'(t)$$

where a,b,c,d are bounded functions $R^+ \to M^n$ and where $u \in D$. The equation is assumed to hold almost everywhere under the assumption that $u'(t)$ denotes the set of right-derived values as explained previously. Existence of $u'(t)$ in the ordinary sense is not required.

If $\|z\| = 1$, $t > 0$ is fixed, and

$$f(z) = a(t) + b(t)z + zd(t) + zc(t)z$$

the tangent condition at t is equivalent to

$$\mathrm{Re}\,\xi^* z^* f(z)\xi \le 0 , \qquad (z^* z\xi = \xi) .$$

With the notation

$$\eta = z\xi , \qquad \xi = z^* \eta \tag{8}$$

this in turn is equivalent to

$$\mathrm{Re}[\eta^* a(t)\xi + \eta^* b(t)\eta + \xi^* d(t)\xi + \xi^* c(t)\eta] \le 0 . \tag{9}$$

If (9) holds for $t > 0$ with ξ and η as in (8), then the Siegel disk is invariant for solutions of (7). In particular, it suffices to have (9) when $|\xi| = |\eta|$; this is an interesting theorem of Reid [3].

If the above condition $[z,f(z)] \le 0$ is required only for unitary z , we get (9) again. Hence, the function $f(z)$ satisfies the tangent condition on the whole boundary of the Siegel disk if it satisfies the tangent condition on the Šilov boundary.

7 MATRIX INEQUALITIES

At a given value of t , the solution u of (7) which satisfies the initial value $u(0) = z$ can be represented as

$$F(z) = A + Bz(1 - Cz)^{-1}D$$

where A,B,C,D are elements of M^n which depend, naturally, on a,b,c,d,t . Invariance of the Siegel disk means $\|F(z)\| \le 1$ for $\|z\| \le 1$. If a,c are permuted, and b,d are permuted, this has a corresponding effect on the coefficients of F , so that the new function is

$$G(z) = C + Dz(1 - Az)^{-1}B \ .$$

Since the suggested permutation has no effect on (9), we are led to surmise that $F(z)$ maps the Siegel disk into itself if, and only if, $G(z)$ does. This is actually true under the hypothesis that $B \neq 0$ and $D \neq 0$. In fact, either condition for F or G is then equivalent to

$$\left\| \begin{pmatrix} \lambda B & C \\ A & D/\lambda \end{pmatrix} \right\| \leq 1 \qquad \text{for some} \quad \lambda > 0 \ ,$$

which is clearly invariant under the permutation described.

As seen in Section 11, results similar to those of Section 6 hold for broad classes of equations $u' = f(t,u)$ where f is a polynomial in (u,u^*) at each t. It is natural to surmise that the solutions of these problems must have symmetry properties similar to those of F and G above, though the subject has apparently not been explored.

8 PHYSICAL INTERPRETATION

Suppose a thin sheet of thickness Δt has transmission and reflection coefficients as shown in Figure 3, apart from a term $o(\Delta t)$ in each case. If this sheet is followed by a medium which has reflection coefficient u, the principal term of the overall reflection $u + \Delta u$ can be obtained by considering the three paths shown in Figure 4. This gives

$$u + \Delta u = a\Delta t + (1 + b\Delta t)u(1 + d\Delta t) + u(c\Delta t)u + o(\Delta t) \ .$$

Cancelling u, dividing by Δt, and letting $\Delta t \to 0$, we get the differential equation (7). Hence, this equation describes

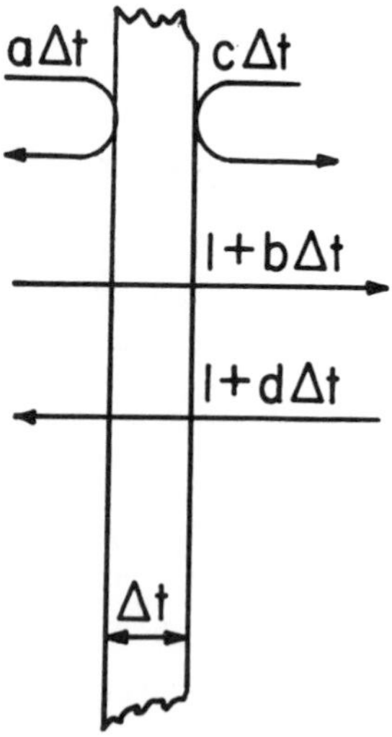

Figure 3

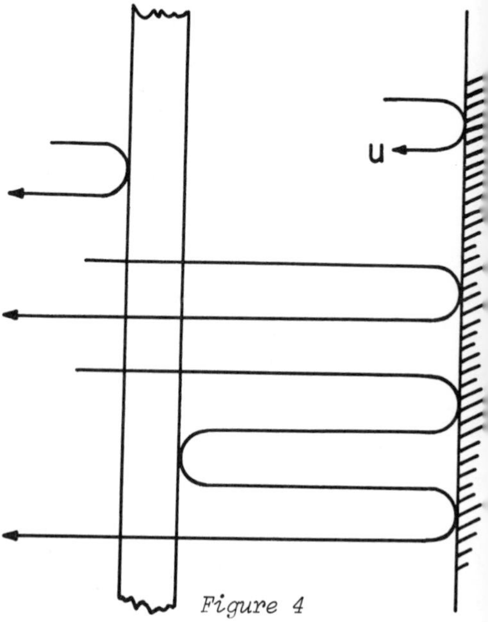

Figure 4

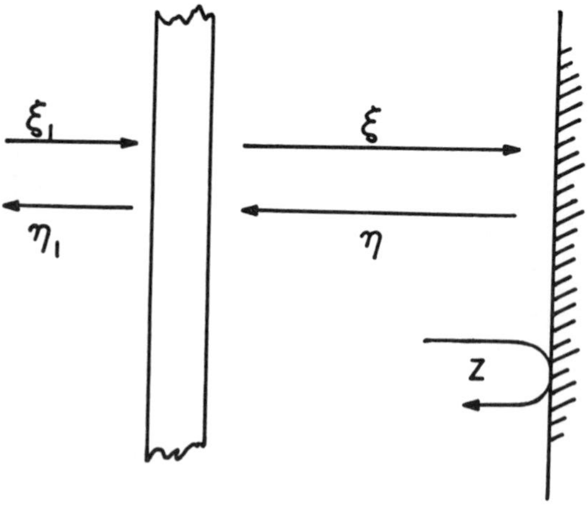

Figure 5

the reflection of a stratified medium as a function of its thickness, t .

Invariance of the Siegel disk for the thin layer means that if $\|z\| \le 1$ in Figure 5, the overall reflection should satisfy the same condition. In other words,

$$\|z\| \le 1 \Rightarrow |\eta_1| \le |\xi_1| .$$

On physical grounds one would surmise that the overall reflection is maximized when the backward wave $\eta = z\xi$ is as large as possible, and this in turn requires $|\eta| = |\xi|$. When $|\eta| = |\xi|$ we could always choose a unitary z for which $\eta = z\xi$; hence, the truly relevant behavior of the waves ξ_1, η_1, ξ, η is not restricted if z is assumed to be unitary. This may serve to show why the tangent condition in (7) is needed only on the Šilov boundary.

For quantitative discussion, note that Figures 3 and 5 give

$$\xi = (1 + b\Delta t)\xi_1 + (c\Delta t)\eta + o(\Delta t)$$

$$\eta_1 = (a\Delta t)\xi_1 + (1 + d\Delta t)\eta + o(\Delta t)$$

or, equivalently,

$$\xi_1 = \xi - (b\xi + c\eta)\Delta t + o(\Delta t)$$

$$\eta_1 = \eta + (a\xi + d\eta)\Delta t + o(\Delta t) .$$

If $|\eta| < |\xi|$ and Δt is small enough, the desired condition $|\eta_1| \le |\xi_1|$ holds without any restriction on a,b,c,d ; this agrees with our surmise that the only case $|\eta| = |\xi|$ need be considered. In that case (9) follows from $|\eta_1| \le |\xi_1|$ for $\Delta t \to 0$, as is easily checked.

9 PHYSICAL INTERPRETATION CONTINUED

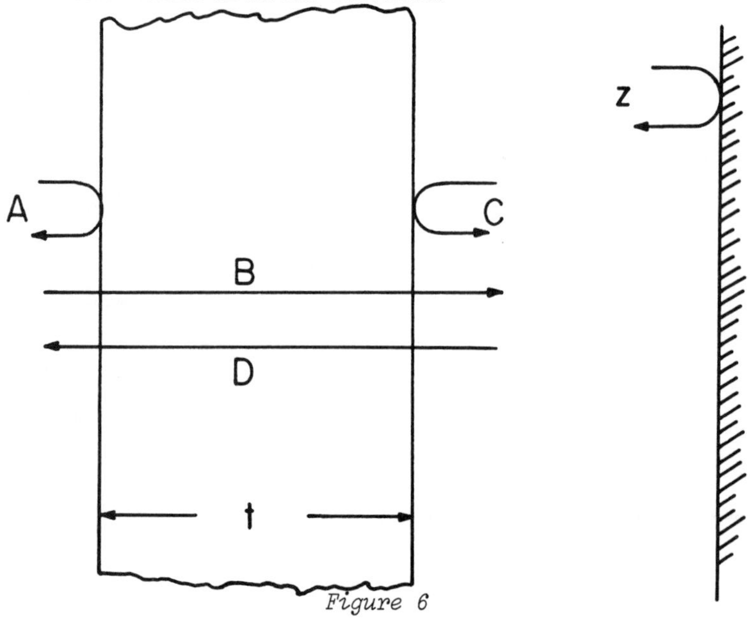

Figure 6

Suppose a sheet of thickness t with reflection and transmission coefficients A,B,C,D is followed by a reflection z as shown in Figure 6. The overall reflection $F(z)$ can be computed as in Figure 4, except that the whole multiple-reflection path must be considered. Thus we get

$$F(z) = A + BzD + BzCzD + BzCzCzD + \cdots$$
$$= A + Bz(1 - Cz)^{-1}D$$

for $\|C\| < 1$, which agrees with the formula for $F(z)$ given previously. The physical configuration is a sheet of thickness t with terminating reflection z . Hence the overall reflection satisfies (7) together with the initial value $u(0) = z$. This interpretation may serve to explain why the solution $u(t,z)$, at fixed t , is expressible as a linear fractional transformation of the initial value, z .

It is possible to introduce vectors ξ and η with $\eta = z\xi$ into Figure 6 similar to those in Figure 5, and deduce that $\|F(z)\|$ attains its maximum when z is unitary. Since $F(z)$ is analytic when $\|C\| < 1$, this agrees with familiar properties of the Šilov boundary.

The condition $\|F(z)\| \le 1$ for $\|z\| \le 1$ expresses the fact that the sheet is passive; that is, has no internal source of energy. Since the condition (9) for passivity of the thin layer is invariant when the layer is turned end for end, one would expect the same invariance to hold for the whole sheet. Turning the sheet end for end has the effect of interchanging A, C and also interchanging B, D; hence, the new reflection is given by the function $G(z)$ introduced above. These remarks give a physical interpretation of the fact that, in general, both functions $F(z)$, $G(z)$ map the Siegel disk into itself if either one does.

10 HILBERT SPACE

We conclude by mentioning some generalizations of the foregoing results that have been obtained in collaboration with Professor Peter Volkmann of Karlsruhe [2]. Let X be a complex Hilbert space and let $L(X)$ denote the set of bounded linear operators on X, with the norms

$$|\xi| = (\xi,\xi)^{\frac{1}{2}}, \qquad \|x\| = \sup_{|\xi|=1} |x\xi|$$

for $\xi \in X$ and $x \in L(X)$, respectively. If $x \in L(X)$ and $y \in L(X)$ we define

$$[x,y] = \lim_{h \to 0+} \sup_{\xi \in \sigma(x,h)} \frac{\mathrm{Re}(x\xi, y\xi)}{\|x\|} \qquad (\|x\| \neq 0)$$

where $\sigma(x,h)$ denotes the set of vectors $\xi \in X$ satisfying

$$|\xi| = 1 , \quad |x^*x\xi - \|x^*x\|\xi| \le h .$$

As in the foregoing discussion, $u'(t)$ denotes the set of right-hand derived values at t, the convergence being in norm; thus, u' and f are functions from R^+ or $R^+ \times L(X)$, respectively, to the power set of $L(X)$. It is assumed (i) that $\|u\|$ is absolutely continuous, or (ii) that $\|u\|$ is continuous and $u'(t)$ is nonempty mod E.

A function $\omega : R^+ \times R^+ \to R^+$ is a *uniqueness function* if $\omega(t,0) = 0$ for $t \in R^+$ and if the conditions $\omega(0) = 0$, $D_+\delta(t) \le \omega(t,\delta(t))$ p.p. imply $\delta = 0$ for functions $\delta : R^+ \to R^+$ satisfying regularity conditions corresponding to (i) or (ii), as the case may be. For example, every Osgood-Tonelli function ω is a uniqueness function for either condition, (i) or (ii). These are functions

$$\omega(t,s) = \alpha(t)\beta(s) , \quad (t \in R^+, s \in R^+)$$

where α is locally integrable, β is continuous, $\alpha \ge 0$, $\beta(s) > 0$ for $s > 0$, and

$$\int_0^1 \frac{ds}{\beta(s)} = \infty .$$

We assume that there exists a uniqueness function ω and a constant $\varepsilon > 0$ such that the conditions

$$1 + \varepsilon > \|x\| \ge 1 , \quad 1 + \varepsilon > \|y\| \ge 1$$

$$\tilde{x} \in f(t,x) , \quad \tilde{y} \in f(t,y)$$

together imply

$$[x - y, \tilde{x} - \tilde{y}] \le \omega(t, \|x - y\|) .$$

Granted all this, the following holds:

THEOREM 3 With X, u and f as above, suppose

(i) Almost all the sets $u'(t) \cap f(t,u(t))$ are nonempty.

(ii) Whenever $\|p\| = 1$ there exists $q \in f(t,p)$ such that $[p,q] \leq 0$.

Then the Siegel disk is invariant.

11 GENERALIZED RICCATI EQUATIONS

If $u(t)$ is a compact operator for almost all t, we can replace $\sigma(x,h)$ by $\sigma(x,0)$ in the definition of $[x,y]$. This sharpens Theorem 2. However, Theorem 3 applies to the Riccati equation and its generalizations with no hypothesis concerning compactness of $u(t)$.

Let us agree that an *admissible monomial* is an expression of form

$$m(t,z) = zz^*zz^* \ldots g(t) \ldots z^*zz^*z$$

where $z \in L(X)$ and $g : \mathbb{R}^+ \to L(X)$ is such that $\|g\|$ is locally integrable. Thus the admissible monomials of degree ≤ 3 are

$$g, zg, gz, zz^*g, zgz, gz^*z, zz^*zg, zz^*gz, zgz^*z, gzz^*z .$$

We then have the following [2]:

THEOREM 4 With X and u as above, let $g(t,z)$ be a finite sum of admissible monomials $m_i(t,z)$. For almost all $t \in R^+$ and all $\xi \in X$ suppose further:

(i) $g(t,u(t)) \in u'(t)$

(ii) $\mathrm{Re}(p\xi, g(t,p)\xi) \leq 0$ whenever p is unitary.

Then the Siegel disk is invariant.

The gist of this result is that, for $g(t,z)$ of the given form, the tangent condition holds on the whole boundary of the Siegel disk if it holds on the Šilov boundary. When applied to the Riccati equation, Theorem 4 extends the theorem of Reid cited in Section 6 to the case in which a,b,c,d,u are functions $R^+ \to L(X)$. This extension can be applied to problems of radiative transfer and astrophysics. For many such problems, it gives global existence of the solution whenever the albedo is less than unity.

REFERENCES

1. Redheffer, R. Matrix differential equations, *Bull. Amer. Math. Soc. 81*(1975), pp. 485-488.

2. Redheffer, R. and P. Volkmann. Invariance properties of operator differential equations, to appear.

3. Reid, W. T. "Riccati Differential Equations." Academic Press, New York, 1972. Additional references can be found here.

LINEAR SECOND ORDER SYSTEMS OF ORDINARY DIFFERENTIAL EQUATIONS AND GEODESIC FLOWS

Philip Hartman

Johns Hopkins University
Baltimore, Maryland

1. INTRODUCTION

This paper is concerned with estimates for solutions of a self-adjoint second order system

(1.1) $$v'' - Q(t)v = 0 \ , \quad \text{or} \quad V'' - Q(t)V = 0 \ ,$$

where $Q(t) = Q^*(t)$ is a continuous, real, symmetric $n \times n$ matrix and $v \in R^n$ or V is an $n \times n$ matrix. The results are suggested by and will be applied (in Section 5) to geodesic flows on Riemann manifolds of negative sectional curvature.

Under the assumption of disconjugacy of (1.1) on $0 \leq t < \omega$ (i.e., no solution $v = v(t) \not\equiv 0$ has two zeros), there exists a principal solution $V = Y(t)$, unique up to constant nonsingular factors on the right; [9] or [4] or [10], pp. 384-396. (For applications of principal solutions to

This study was supported by NSF Grant MPS75-15733.

ISBN 0-12-632250-3

differential geometry, see for example [14], [5] and references in Section 5 below.) If $\omega = \infty$ and $Q \geqq 0$, then $Y(t)$ can be constructed as follows: Let $V = X(t)$ be the solution of (1.1) determined by $X(0) = I$, $X'(0) = cI$ with $c \geqq 0$, then

$$Y(t) = X(t) \int_t^\infty X^{-1}(s) X^{*-1}(s) ds . \tag{1.2}$$

As will be seen below, if $Q = Q^* \geqq k^2 I > 0$ for some constant $k > 0$ and $c \geqq k$, then $|X^{-1}(t)| \leqq e^{-kt}$ and $|Y(t)| \leqq e^{-kt}/k$. When (1.1) is a scalar equation, the arguments of [8], p. 573, involving L'Hôpital's rule, show that $0 < X(t)Y(t) \leqq 1/2k$ for $t \geqq 0$. This suggests the question of estimates for $|X(t)Y(t)|$ or $|X(t)| \cdot |Y(t)|$ in the matrix case. A related question concerns estimates for $|X(t)| \cdot |X^{-1}(t)|$.

Generally, results will be obtained from the 1-dimensional case. In Section 2, we state or derive the desired results for the scalar case. Section 3 involves Rauch type of comparison theorems where we assume $Q(t) \geqq q(t)I$ and/or $Q(t) \leqq q(t)I$ for real-valued $q(t)$. In this section, we also obtain estimates of the form $|X(t)| \cdot |X^{-1}(t)|^{1/\theta} \leqq C$. Section 4 concerns principal solutions: estimates, comparison theorems, and smooth dependence on parameters. In Section 5, we apply the results to the question of the smoothness of the Anosov decomposition of the tangent bundle $T(SM)$ for the geodesic flow on the unit tangent bundle SM of a Riemann manifold M with negative sectional curvature.

Estimates for solutions will generally be obtained by first deriving an inequality for solutions of related Riccati equations and then applying a quadrature. Actually, the required inequalities for Riccati equations are known for much more general situations; cf., e.g., Coppel [4], pp. 49-58. But, in eneral, these inequalities cannot be integrated to give inequal-ties for solutions of (1.1); cf. the familiar case of the general turm inequalities for two equations of the form $(pv')' - qv = 0$. 'or this reason, we restrict our considerations to an equation of he form (1.1), and compare it with a scalar equation. In view of the simplicity of the proofs and for the sake of completeness, we shall frequently derive the required inequality for Riccati equations from the scalar case, instead of using more general results. Cf. also [0].

2. SCALAR CASE

For ready reference below, we recall some simple facts about second order differential inequalities, and derive some new ones.

PROPOSITION 2.1 Let $q(t) \in C^0[0,\omega)$, $0 < \omega \leqq \infty$, and let $x(t)$ and $y(t)$ be solutions of the differential inequalities

$$x'' \leqq q(t)x \quad \text{and} \quad y'' \geqq q(t)y \tag{2.1}$$

on $0 \leqq t < \omega$, positive on $0 < t < \omega$. (i) If the inequality $x'/x \leqq y'/y$ holds at $t = 0$ (in the sense that $(\infty \geqq)$ $\limsup(y'x - yx') \geqq 0$ as $t \to +0$), then it holds for $0 < t < \omega$ and so, $0 < x(t)/x(s) \leqq y(t)/y(s)$ for $0 < s \leqq t$. (ii) If $\omega = \infty$, and $x > 0$, $x' \leqq 0$ and $y > 0$, $y' \leqq 0$ for $t \geqq 0$, then $0 \geqq x'/x \geqq y'/y$ and $x(t)/x(s) \geqq y(t)/y(s)$ for $0 < s \leqq t$.

Part (i) is an immediate consequence of $y'x - yx' \geqq 0$. Part (ii) is contained in a comparison theorem of Wintner and myself; cf. [10], Corollary 6.5, pp. 358-359.

Below we shall need comparison theorems for certain solutions of the differential inequalities

(2.2) $x'' \leqq q(t)x$ and $y'' \geqq \delta^2 q(t)y$, where $0 < \delta < 1$

and δ is a constant.

PROPOSITION 2.2 Let $q(t) \in C^0[0,\infty)$ and let $x(t), y(t)$ be positive solutions of (2.2) satisfying

$$x'/x \leqq y'/\delta^2 y \tag{2.3}$$

at $t = 0$. Then (2.3) holds for $t \geqq 0$ and so,

$$0 < x(t)/x(s) \leqq [y(t)/y(s)]^{1/\delta^2} \quad \text{for} \quad t \geqq s \geqq 0 \ . \tag{2.4}$$

If $q(t)$ is a positive constant, then δ^2 in (2.3) and (2.4) can be replaced by δ . In order to obtain inequalities better than (2.3) and (2.4), we shall impose additional conditions on $q(t)$ and $y(t)$.

PROPOSITION 2.3 Let $q(t)$, $x(t)$, $y(t)$, δ be as in Proposition 2.2. Suppose, in addition, that k, K, θ are positive constants satisfying

$$\begin{gathered} 0 < k^2 \leqq q(t) \leqq K^2 \ , \\ k^2/K^2 \geqq \theta(\theta - \delta^2)/(1 - \theta) \ , \quad \delta^2 \leqq \theta < 1 \ , \end{gathered} \tag{2.5}$$

$$y'/y \geqq k \quad \text{for} \quad t \geqq 0 \ . \tag{2.6}$$

If the inequality

$$x'/x \leqq y'/\theta y \tag{2.7}$$

holds at $t = 0$, then it holds for $t \geqq 0$ and so,

$$0 < x(t)/x(s) \leqq [y(t)/y(s)]^{1/\theta} \quad \text{for} \quad t \geqq s > 0 \ . \tag{2.8}$$

PROOFS OF PROPOSITIONS 2.2 AND 2.3 For any $\theta > 0$, let $y = u^\theta$, so that $u'/u = y'/\theta y$. Then u satisfies the ifferential inequality

$$u'' \geqq [(\delta^2/\theta)q(t) + (1 - \theta)(u'/u)^2]u \equiv q_1(t)u \ .$$

ote that the coefficient function q_1 satisfies $q_1 \geqq q$ if $= \delta^2$. In this case, Proposition 2.2 follows from Proposition .1.

If one assumes (2.6), then

$$q_1 \geqq (\delta^2/\theta)q + (1 - \theta)k^2/\theta^2$$

2.9)
$$= q(t) + \{(1 - \theta)k^2/\theta^2 - (1 - \delta^2/\theta)q\} \ .$$

ence if $\{\ldots\} \geqq 0$, then $q_1 \geqq q$ holds and Proposition 2.1 mplies (2.7) and (2.8). In view of the first part of (2.5), his is the case if $(1 - \theta)k^2/\theta^2 \geqq (1 - \delta^2/\theta)K^2 \geqq$ $1 - \delta^2/\theta)q(t)$, i.e., if θ satisfies the inequalities in 2.5). This completes the proof.

The assumption (2.6) appears artificial but arises in an pplication below where $\delta^2q(t)$ in (2.2) can be replaced by ax$(\delta^2q(t),k^2)$. Under different circumstances, it might be ore reasonable to assume that $y'/y \geqq \delta k$ holds at $t = 0$. n this case, we have

PROPOSITION 2.4 Let $q(t)$, $x(t)$, $y(t)$, δ be as in roposition 2.2. Let k,K,θ be positive constants satisfying

2.10)
$$0 < k^2 \leqq q(t) \leqq K^2 \ ,$$
$$k^2/K^2 \geqq \theta(\theta - \delta^2)/\delta^2(1 - \theta) \ , \quad \delta^2 \leqq \theta < 1 \ ,$$

nd let $y'/y \geqq \delta k$ hold at $t = 0$. If (2.7) holds at $t = 0$, hen it holds for $t \geqq 0$ and (2.8) holds.

PROOF By Proposition 2.1, $y'/y \geqq \delta k$ holds for $t \geqq 0$. n the proof above, the term $(1 - \theta)k^2/\theta^2$ must be replaced by

$(1 - \theta)k^2\delta^2/\theta^2$, and the proof can be completed as before.

PROPOSITION 2.5 Let $q(t)$, $x(t)$, $y(t)$, δ be as in Proposition 2.2. (i) If $q > 0$ is nonincreasing and the inequalities $0 < q^{\frac{1}{2}} \leqq y'/\delta y$ and $x'/x \leqq y'/\delta y$ hold at $t = 0$, then they hold for $t \geqq 0$, so that (2.7) holds with $\theta = \delta$. (ii) If $q > 0$ is nondecreasing and the inequality $x'(t)/x(t) \leqq y'(t/\delta)/\delta y(t/\delta)$ holds at $t = 0$, then it holds for $t \geqq 0$, so that $x(t)/x(s) \leqq y(t/\delta)/y(s/\delta)$ for $t \geqq s \geqq 0$.

PROOF On (i). Suppose that $q > 0$ is nonincreasing. By Proposition 2.1, we can suppose that $y'' = \delta^2 qy$, so that $\rho = y'/y > 0$ satisfies $\rho' = \delta^2 q - \rho^2$. Hence $\sigma = \rho/\delta$ satisfies $\sigma' = \delta(q - \sigma^2)$ and $\sigma^2(0) \geqq q(0) > 0$. Since q is nonincreasing, it follows that $\sigma^2 \geqq q$ for $t \geqq 0$ (for if $t_0 > 0$, $0 < \sigma^2(t_0) < q(t_0)$, then $\sigma' \geqq$ const. > 0 on any interval $[t, t_0]$ on which $\sigma > 0$; cf. [11], e.g., [10], Exercise 3.9(d), p. 515 and p. 579). Hence $q - \sigma^2 \leqq 0$ for $t \geqq 0$ and $\sigma' = \delta(q - \sigma^2) \geqq q - \sigma^2$. Since $r = x'/x$ satisfies $r' \leqq q - r^2$ and $r(0) \leqq \sigma(0)$, (i) follows.

On (ii). Suppose that $q > 0$ is nondecreasing. If $\sigma = \rho/\delta$ and $\rho = y'/y$, then $\sigma' \geqq \delta(q - \sigma^2)$. Thus $\tau = \sigma(t/\delta)$ satisfies, $\tau' \geqq q(t/\delta) - \tau^2$ or, since q is nondecreasing, $\tau' \geqq q(t) - \tau^2$. Thus (ii) follows from $\tau(0) \geqq r(0)$, $r = x'/x$, $r' \leqq q - r^2$.

3. RAUCH TYPE COMPARISON THEOREMS

We prove Propositions 3.1 and 3.2, which imply Rauch's comparison theorem ([16], Theorem 4, p. 43). These results are educed rather simply from Sturm's comparison theorem as given in roposition 2.1 above.

If $A = A^*$ and $B = B^*$ are (real) symmetric matrices, we se the standard notation $A \geqq B$ to mean that the corresponding uadratic forms satisfy $(A\xi,\xi) \geqq (B\xi,\xi)$ for all vectors ξ .

PROPOSITION 3.1 Let $Q(t) = Q^*(t)$ be a continuous real ymmetric matrix function on $0 \leqq t < \omega (\leqq \infty)$ and let $v = v(t)$ e a vector solution of

$$v'' - Q(t)v = 0 , \tag{3.1}$$

$(t) \neq 0$ on $0 < t < \omega$. Then $y = |v(t)|$ satisfies

$$y'' \geqq q(t)y \tag{3.2}$$

f $q = (Qv,v)/|v|^2$ or, say

$$Q(t) \geqq q(t)I , \tag{3.3}$$

.g., $q(t)$ is the smallest eigenvalue of $Q(t)$. Hence roposition 2.1 is applicable if $x(t)$ is a suitable solution f (2.1).

PROOF From $y^2 = |v(t)|^2 = (v,v)$, it is seen that $y' = (v',v)$, so that $|y'(t)| \leqq |v'(t)|$. Also, $yy'' + y'^2 =$ $Qv,v) + |v'|^2$, and so $yy'' \geqq (Qv,v) \geqq qy^2$. This gives (3.2).

We shall need the following standard fact: If $Q = Q^*$ and $,W$ are (square) matrix solutions of

$$V'' - Q(t)V = 0 , \tag{3.4}$$

hen the "Wronskian" $W^*V' - W^{*\prime}V$ is a constant. In articular, for $W = V$, $V^*V' - V^{*\prime}V$ is constant. When this onstant is zero, $V^*V' - V^{*\prime}V = 0$, i.e., $(V^*V') = (V^*V')^*$, we shall say the solution $V(t)$ is *prepared*. If V is nonsingular, this is equivalent to $V'V^{-1} = V^{*-1}V^{*\prime}$, i.e., $(V'V^{-1}) = (V'V^{-1})^{*}$.

PROPOSITION 3.2 Let $Q(t) = Q^{*}(t)$ be a continuous real symmetric function on $0 \leqq t < \omega(\leqq\infty)$. Let $q(t) \in C^{0}[0,\omega)$ and

$$(3.5) \qquad Q(t) \leqq q(t)I$$

(e.g., $q(t)$ is the largest eigenvalue of $Q(t)$). Let $V(t)$ be a prepared solution of (3.4), nonsingular for $0 < t < \omega$, and $c = \liminf |V'(t)V^{-1}(t)|(\leqq\infty)$ as $t \to +0$. Then the differentia inequality (3.2) has a solution $y(t)$, positive on $0 < t < \omega$, and $(\infty\geqq) \limsup y'(t)/y(t) \geqq c$ as $t \to +0$. Furthermore, for any such $y(t)$,

$$(3.6) \qquad V'V^{-1} \leqq (y'/y)I , \quad \text{so that} \quad (V'\xi,V\xi)/|V\xi|^{2} \leqq y'/y ,$$

$$(3.7) \qquad |V\xi|' \leqq (y'/y)|V\xi| ,$$

$$(3.8) \qquad |V(t)\xi| \leqq [y(t)/y(s)] \cdot |V(s)\xi| \quad \text{and} \quad |V(s)V^{-1}(t)| \leqq y(s)/y(t) \quad \text{for} \quad 0 \leqq s \leqq t .$$

If $Q(t) \geqq 0$ and $V'(0)V^{-1}(0) > 0$, then $V'V^{-1} > 0$ for $0 \leqq t < \omega$ and (3.7) can be improved to $|V\xi|' \leqq |V'\xi| \leqq (y'/y)|V\xi|$.

The inequality (3.6) is contained in more general results; cf. Coppel [4], pp. 49-58.

PROOF The symmetric matrix $U = V'V^{-1}$ satisfies the Riccati equation $U' = Q - U^{2}$. Let $|\xi| = 1$ and $r = (U\xi,\xi)$, so that $r^{2} \leqq |U\xi|^{2} = (U^{2}\xi,\xi)$. Hence $r' \leqq q(t) - r^{2}$. Assume, for a moment, the existence of a $y(t)$ as above. Then, since

$\rho = y'/y$ satisfies $\rho' \geqq q - \rho^2$ and $\liminf r(t) \leqq c \leqq \limsup \rho(t)$ as $t \to +0$, it follows that $r(t) \leqq \rho(t)$. This gives the first inequality in (3.6), i.e., $(V'V^{-1}\xi,\xi) \leqq (y'/y)|\xi|^2$. On replacing ξ by $V(t)\xi$, we get the second inequality in (3.6). The latter is equivalent to (3.7). Finally (3.8) follows from a quadrature.

Actually, this argument shows the existence of a $y(t)$, for example, satisfying the equation $y'' = q(t)y$; cf. [10], Theorem 7.2, p. 362. This completes the proof of the first part.

In order to verify the last part, note that if $Q \geqq 0$, then $|V(t)\xi|$, hence $|V(t)\xi|^2$, is a convex function of t, by Proposition 3.1 with $q(t) = 0$ in (3.2). Thus $(V'(t)\xi,V(t)\xi) \geqq (V'(0)\xi,V(0)\xi) = (V'(0)V^{-1}(0)\eta,\eta) > 0$ if $\eta = V(0)\xi$. Hence $(U(t)\xi,\xi) = (V'(t)V^{-1}(t)\xi,\xi) > 0$. Consequently, $0 < U \leqq (y'/y)I$, and so $|U| \leqq y'/y$. This gives the desired result.

COROLLARY 3.1 Let $Q_1(t),Q_2(t)$ be continuous, real symmetric matrix functions on $0 \leqq t < \omega$. Let $q(t) \in C^0[0,\omega)$ satisfy $Q_1 \leqq qI \leqq Q_2$. Let $V_j(t)$ be a prepared solution of $V_j'' - Q_jV_j = 0$ for $j = 1,2$, such that $V_1(t)$ is nonsingular for $0 < t < \omega$, $V_2(t)$ is nonsingular for small $t > 0$ and $\liminf|V_1'V_1^{-1}| \leqq \limsup|V_2'V_2^{-1}|(\leqq\infty)$ as $t \to +0$. Then $V_2(t)$ is nonsingular for $0 < t < \omega$ and $|V_1(t)\xi| \leqq |V_2(t)\eta|$ if $|\xi| = |\eta| = 1$; in particular, $|V_1| \leqq |V_2|$ and $|V_2^{-1}| \leqq |V_1^{-1}|$.

Corollary 3.1 is a consequence of Propositions 3.1 and 3.2. It contains Rauch's [16] comparison theorem and Berger's [3] variant of it.

It will be convenient to record for use below particular cases of Propositions 3.1 and 3.2, where $q(t)$ is constant.

COROLLARY 3.2 Let $Q(t) = Q^*(t) \in C^0[0,\infty)$ and $Q(t) \geqq k^2 I$, where $k > 0$ is a constant. Let $V = X(t)$ be the solution of (3.4) determined by

$$X(0) = I, \quad X'(0) = cI \tag{3.9}$$

with $0 < k \leqq c$. Then $X(t)$ is prepared, X' and X^{-1} are nonsingular for $t \geqq 0$, $X'X^{-1} \geqq kI$ and, for $\xi \neq 0$,

$$|X'(t)\xi| \geqq |X(t)\xi|' \geqq k|X(t)\xi|, \tag{3.10}$$

$$|X(s)X^{-1}(t)| \leqq e^{-k(t-s)} \quad \text{for} \quad t \geqq s \geqq 0. \tag{3.11}$$

This follows from Proposition 3.1 with $q(t) = k^2$, $y(t) = |X(t)\xi|$, $y(0) = |\xi|$, $y'(0) = c|\xi|$ and $x(t) = e^{kt}$.

COROLLARY 3.3 Let $Q(t) = Q^*(t) \in C^0[0,\infty)$ and $0 \leqq Q \leqq K^2 I$, where $K > 0$ is a constant. Let $V = X(t)$ be the solution of (3.4) satisfying (3.9) with $0 < c \leqq K$. Then X', X^{-1} are nonsingular for $t \geqq 0$, $0 < X'X^{-1} \leqq KI$ and, for $\xi \neq 0$,

$$|X(t)\xi|' \leqq |X'(t)\xi| \leqq K|X(t)\xi|, \tag{3.12}$$

$$\begin{aligned} &|X(t)\xi| \leqq |X(s)\xi|e^{K(t-s)}, \\ \text{so} \quad &|X(s)X^{-1}(t)| \geqq e^{-K(t-s)} \quad \text{for} \quad t \geqq s \geqq 0. \end{aligned} \tag{3.13}$$

This follows from Proposition 3.2 with $q(t) = K^2$ and $y(t) = e^{Kt}$.

THEOREM 3.1 Let $Q(t) = Q^*(t) \in C^0[0,\infty)$ satisfy $p(t)I \leqq Q(t) \leqq q(t)I$, where $0 < k^2 \leqq p(t) \leqq q(t) \leqq K^2$; $p, q \in C^0[0,\infty)$; and $0 < k < K$ are constants. Let $V = X(t)$ be the solution of (3.4) satisfying (3.9) with $k \leqq c \leqq K$.

hen (3.10)-(3.13) hold. If, in addition, it is supposed that (t)/q(t) $\geqq \delta^2$, where $0 < \delta < 1$ and that θ, k, K satisfy 2.5), then $|X(t)| \cdot |X^{-1}(t)|^{1/\theta} \leqq 1$. (The conditions in (2.5) re redundant if $\theta = \delta^2$.)

PROOF Let $|\xi| = |\eta| = 1$. By Proposition 3.1, the func- ion $y = |X(t)\xi|$ satisfies $y'' \geqq p(t)y \geqq \delta^2 q(t)y$ and $y(0) =$, $y'(0) = c$. Also, by Proposition 3.2, $z = |x(t)\eta|$ atisfies $z'/z \leqq x'/x$, where $x'' \leqq q(t)x$ and $x(0) = 1$, '(0) = c . It follows from Proposition 2.3 and Corollary 3.2 hat $z(t) \leqq x(t) \leqq y(t)^{1/\theta}$, i.e., $|X(t)\xi| \leqq |X(t)\eta|^{1/\theta}$. If is replaced by $X^{-1}(t)\eta/|X^{-1}(t)\eta|$, we obtain $X(t)\xi| \cdot |X^{-1}(t)\eta|^{1/\theta} \leqq 1$. This completes the proof.

. PRINCIPAL SOLUTIONS

When $P = P^* > 0$, $Q = Q^*$ and

$$(PV')' - QV = 0 \tag{4.1}$$

s disconjugate on $0 \leqq t < \omega$, then (4.1) has a principal olution $V = Y(t)$, see [9] or [10], pp. 393-396 or [5]; for eneralizations, see Reid [17]. A principal solution Y is a repared (i.e., $Y^*PY' = Y'^*PY$), nonsingular solution, unique up o multiplication by nonsingular constant matrices on the right, atisfying

$$\min_{|\xi|=1} \int_0^T (PY^{-1}\xi, Y^{-1}\xi)ds \to \infty \quad \text{as} \quad T \to \omega .$$

t is also given by $Y(t) = \lim Y_T(t)$ as $T \to \omega$, where $= Y_T(t)$ is the solution of (4.1) determined by the boundary onditions $Y_T(0) = I$, $Y_T(T) = 0$ and "lim" indicates a C^2- imit on every $[0,s]$, $0 < s < \omega$. When $P = I$, $Q \geqq 0$ and

$\omega = \infty$, a principal solution Y is also characterized by $|Y(t)\xi| > 0$, $|Y(t)\xi|' \leqq 0$ for $0 \leqq t < \omega$ and $\xi \neq 0$. Proposition 4.2 can be considered as an extension of a comparison theorem for principal solutions of scalar equations (Hartman and Wintner; cf. [10], Corollary 6.5, pp. 358-359). For a more general result, see Coppel [4], Corollary, p. 54.

PROPOSITION 4.1 Let $P = P^*$, $Q = Q^*$ be continuous for $0 \leqq t < \omega$, and $P > 0$, $Q \geqq 0$. Then a principal solution $V = Y(t)$ of (4.1) satisfies $PY'Y^{-1} \leqq 0$ for $0 \leqq t < \omega$.

PROOF For by $(V^*PV')' = V^{*}{}'PV' + V^*QV \geqq 0$, the solution $V = Y_T(t)$ of (4.1) satisfies

$$Y_T^*PY_T'(t) = - \int_t^T (Y_T^*PY_T' + Y_T^*QY_T)ds \leqq 0 \quad \text{for} \quad 0 \leqq t \leqq T .$$

Hence $Y^*PY' \leqq 0$ for $0 \leqq t < \omega$ and the proposition follows, since Y is nonsingular.

PROPOSITION 4.2 Let $Q_j(t) = Q_j^*(t)$ be continuous for $0 \leqq t < \omega$ and $j = 1,2$, $Q_1 \leqq Q_2$, and let

$$(4.2_j) \qquad V'' = Q_j(t)V ,$$

$j = 1$, be disconjugate. Then (4.2_2) is disconjugate on $0 \leqq t < \omega$. Let $V = Y_j(t)$ be a principal solution of (4.2_j). Then $Y_1'Y_1^{-1} \geqq Y_2'Y_2^{-1}$ for $0 \leqq t < \omega$.

The proof is similar to the Case 1 of the proof of [10], Corollary 6.5, pp. 358-359, for the variation of constants $V = Y_1Y$ reduces (4.2_2) to (4.1) with $P = Y_1^*Y_1 > 0$, $Q = Y_1^*(Q_2 - Q_1)Y_1 \geqq 0$. If Y is a principal solution of (4.1), then $Y_2 = Y_1Y$ is a principal solution of (4.2_2), and

$$Y_2'Y_2^{-1} = Y_1'Y_1^{-1} + Y_1^*(PY'Y^{-1})Y_1 \leqq Y_1'Y_1^{-1} ,$$

y Proposition 4.1. This has the following corollaries if ither $Q_1 = qI$ or $Q_2 = qI$.

COROLLARY 4.1 Let $Q(t) = Q^*(t) \in C^0[0,\omega)$ satisfy $\geqq qI$, where $q \in C^0[0,\omega)$ and

$$(4.3) \qquad y'' = q(t)y$$

s disconjugate on $[0,\omega)$. Then (3.1) is disconjugate on $0,\omega)$. If $Y(t)$ and $y(t)$ are principal solutions of (3.4) nd (4.3), then

$$(4.4) \qquad Y'Y^{-1} \leqq (y'/y)I \text{ , so that } (Y'\xi,Y\xi)/|Y\xi|^2 \leqq y'/y$$

or $\xi \neq 0$, i.e., $|Y\xi|'/|Y\xi| \leqq y'/y$ and, for $0 \leqq s \leqq t$,

$$(4.5) \qquad \begin{aligned} &|Y(t)\xi|/|Y(s)\xi| \leqq y(t)/y(s) \\ \text{and } &|Y(t)Y^{-1}(s)| \leqq y(t)/y(s) , \end{aligned}$$

$$(4.6) \qquad \int^\omega |Y(t)\xi|^{-2}dt = \infty \quad \text{for } \xi \neq 0 .$$

Proposition 4.2 implies the first part of (4.4) which, in urn, implies the last part of (4.4) and (4.5). Also, (4.6) ollows from $\int^\omega |y(t)|^{-2}dt = \infty$; Leighton and Morse, cf. [10], heorem 6.4, p. 355.

In [9], pp. 29-30, we raised the question whether a rincipal solution of (3.4) can be characterized by (4.6), as in he case $n = 1$. Corollary 4.1 answers this in the affirmative hen $Q \geqq qI$ and (4.3) is disconjugate.

COROLLARY 4.2 Let $Q = Q^* \in C^0[0,\omega)$ satisfy $Q \leqq qI$, here $q \in C^0[0,\omega)$. Let (3.1) be disconjugate on $[0,\omega)$.

Then (4.3) is disconjugate on $[0,\omega)$ and if $Y(t)$ and $y(t)$ are principal solutions of (3.4) and (4.3), then

$$Y'Y^{-1} \geqq (y'/y)I \,, \quad \text{so that} \quad (Y'\xi, Y\xi)/|Y\xi|^2 \geqq y'/y \tag{4.7}$$

for $\xi \neq 0$, so that $|Y\xi|'/|Y\xi| \geqq y'/y$ and

$$|Y(t)\xi| \geqq |Y(s)\xi|y(t)/y(s) \quad \text{and} \quad |Y(s)Y^{-1}(t)| \leqq y(s)/y(t) \quad \text{for} \quad 0 \leqq s \leqq t \,. \tag{4.8}$$

PROPOSITION 4.3 Let $Q = Q^* \in C^0[0,\infty)$, $Q \geqq k^2 I$, where $k \geqq 0$. Let $V = X(t)$ be the solution of (3.4) satisfying (3.9 with $c \geqq k$. Then

$$Y(t) = X(t) \int_t^\infty X^{-1}(s) X^{*-1}(s)\,ds \tag{4.9}$$

is a principal solution ([9] or [10], p. 393) and

$$|Y(t)\xi|' \leqq -k|Y(t)\xi| \leqq 0 \,, \quad |Y(t)\xi| \leqq |Y(s)\xi| e^{-k(t-s)} \quad \text{for} \quad 0 \leqq s \leqq t \,. \tag{4.10}$$

If $k > 0$, then

$$|X^{-1}(t)Y(t)X^*(t)| \leqq \int_t^\infty |X^{-1}(s)|\,ds \leqq e^{-kt}/k \,. \tag{4.11}$$

PROOF The inequalities (4.10) follow from Corollary 4.1. In order to verify (4.11), note that, by (3.11), the left side is at most

$$\int_t^\infty |X^{-1}(s)| \cdot |X^{*-1}(s)X^*(t)|\,ds \leqq \int_t^\infty |X^{-1}(s)|\,ds \,.$$

THEOREM 4.1 Let $Q(t,\alpha) = Q^*(t,\alpha) \in C^0([0,\infty) \times [0,1])$ atisfy $Q \geqq 0$, $Q_\alpha = \partial Q/\partial\alpha$ exists and is continuous in (t,α). et $X(t)$, $Y(t)$ be as in the first part of Proposition 4.3. hen a sufficient condition for $Y(t,\alpha)$, $Y'(t,\alpha)$ to be of lass C^1 is that

$$\int_0^\infty |X^{-1}(t)|\{\int_0^t |Y(s)| \cdot |Q_\alpha(s)|ds$$

$$+ |X^{-1}(t)Y(t)X^*(t)| \int_0^t |Q_\alpha(s)|ds\}dt \tag{4.12}$$

onverge uniformly for $0 \leqq \alpha \leqq 1$.

PROOF It is clear that X,X' are of class C^1 . Also ,Y' are of class C^1 if and only if $Y(0,\alpha),Y'(0,\alpha)$ are of lass C^1 . This is the case if and only if

$$Y(0,\alpha) = \int_0^\infty X^{-1}(s)X^{*-1}(s)ds = (Y'(0,\alpha) + I)/c$$

s of class C^1 . Formally,

$$\partial Y(0,\alpha)/\partial\alpha = - \int_0^\infty (X^{-1}X_\alpha X^{-1}X^{*-1} + X^{-1}X^{*-1}X_\alpha^* X^{*-1})ds .$$

his makes it clear that Y,Y' are of class C^1 if

$$\int_0^\infty |X^{-1}X_\alpha X^{-1}| \cdot |X^{-1}|ds \text{ is uniformly convergent} \tag{4.13}$$

or $0 \leqq \alpha \leqq 1$. We have $X_\alpha'' - QX_\alpha = Q_\alpha X$ and $X_\alpha(0) =$ $X_\alpha'(0) = 0$. Since (4.9) implies that $XY^* - YX^* = 0$ and $X'Y^* - Y'X^* = I$,

$$X_\alpha(t) = X(t) \int_0^t Y^*(s)Q_\alpha(s)X(s)ds - Y(t) \int_0^t X^*(s)Q_\alpha(s)X(s)ds .$$

Hence $X^{-1}(t)X_\alpha(t)X^{-1}(t)$ is

$$\int_0^t Y^*(s)Q_\alpha(s)[X(s)X^{-1}(t)]ds$$

$$- X^{-1}(t)Y(t)X^*(t)\int_0^t [X^{*-1}(t)X^*(s)]Q_\alpha(s)[X(s)X^{-1}(t)]ds ,$$

and so, its norm does not exceed

$$\int_0^t |Y(s)| \cdot |Q_\alpha(s)|ds + |X^{-1}(t)Y(t)X^*(t)| \int_0^t |Q_\alpha(s)|ds ,$$

by (3.11). Hence Theorem 4.1 follows.

For use below, we state the following:

COROLLARY 4.1 If, in Theorem 4.1, there exist constants $k > 0$, $\theta > \frac{1}{2}$, $C > 0$ such that $Q \geqq k^2 I$, $|Q_\alpha| \leqq C|X|$, and $|X| \cdot |X^{-1}|^{1/\theta} \leqq C$, then $Y(t,\alpha), Y'(t,\alpha)$ are of class C^1 .

REMARK Note that the assumption $|X| \cdot |X^{-1}|^{1/\theta} \leqq C$, $\theta > \frac{1}{2}$, is always satisfied with $\theta = 1$ if $\dim v = 1$. For systems of arbitrary dimension, Theorem 3.1 gives sufficient conditions for this assumption.

PROOF By (4.11) and Fubini's theorem, the integral $(\leqq \infty)$ in (4.12) is at most

$$\int^\infty |Y(s)| \cdot |Q_\alpha(s)| \int_s^\infty |X^{-1}(t)|dtds$$

$$+ \int^\infty |Q_\alpha(s)| \left(\int^\infty |X^{-1}(t)|dt\right)^2 ds .$$

Since (3.11) implies that

$$|Y(t)| \leq \int_t^\infty |X(t)X^{-1}(s)| \cdot |X^{-1}(s)|ds \leq \int_t^\infty |X^{-1}(s)|ds$$

and since $|Q_\alpha| \leq C|X| \leq C|X^{-1}|^{-1/\theta}$, the integral in (4.12) is majorized by a constant times

$$\int_0^\infty |X^{-1}(s)|^{-1/\theta}\left(\int_s^\infty |X^{-1}(t)|dt\right)^2 ds$$

$$\leq \int_0^\infty \left(\int_s^\infty |X^{-1}(t)|^{1-1/2\theta}dt\right)^2 ds$$

$$\leq \int_0^\infty \left(\int_s^\infty e^{-k(1-1/2\theta)t}dt\right)^2 ds < \infty ,$$

since $1 - 1/2\theta > 0$. This proves the corollary.

5. GEODESIC FLOWS

Let $M = M^{n+1}$ be a complete smooth Riemann manifold of dimension $n + 1$ with negative sectional curvature $-K_x(\pi) \geq k^2 > 0$, where $K_x(\pi)$ is the sectional curvature of a 2-plane $\pi \subset T_x(M)$, $x \in M$. Let SM be the bundle of unit tangent vectors, $SM = \{(x,x') : x \in M , x' \in T_x(M) , \|x'\| = 1\}$. The geodesic flow φ^t on SM is defined by $\varphi^t(x(0),x'(0)) = (x(t),x'(t))$, where $x(t)$ moves along a geodesic with initial point $(x(0),x'(0))$. The corresponding linearized flow $d\varphi^t$ on $T(SM)$ is given by $(x(0),x'(0),z(0),z^1(0)) \rightarrow (x(t),x'(t),z(t),D_Tz(t))$, where $z(t)$ is the solution of the Jacobi equation along the geodesic $x = x(t)$,

(5.1) $$D_T^2 z + R(z,T)T = 0 ,$$

with initial condition $z(0)$, $(D_T z)(0) = z^1(0)$, T is the tangent vector $x'(t)$, D_T denotes covariant differentiation with respect to the vector field $T = x'(t)$, and $R = R(x(t))$ is the Riemann curvature tensor at $x = x(t)$.

As is known [1] (also [2]), and as will be seen below, $-K_x(\pi) \geqq k^2 > 0$ implies that the tangent bundle $T(SM)$ has a decomposition as a sum of continuous bundles, invariant under $d\varphi^t : T(SM) = E^s + E^0 + E^u$, where the fibres of E^0 are 1-dimensional and consist of vectors tangent to the flow φ^t , E^s [or E^u] is the "stable" [or "unstable"] n-dimensional bundle, so that $\|z(t)\| + \|D_T z\| \leqq (\|z(0)\| + \|D_T z(0)\|)e^{-kt}$ for $t > 0$ [or $t < 0$] if $(z(0), D_T z(0)) \in E^s_{(x(0),x'(0))}$ [or $E^u_{(x(0),x'(0))}$] . That is, if M is compact, then φ^t is an Anosov (= hyperbolic) flow on SM .

As observed in [1], in general, in the Anosov decomposition of $T(N) = E^s + E^0 + E^u$ for a smooth Anosov flow ψ^t on a smooth manifold N , the bundles E^s, E^u need not be of class C^1 . For the case $N = SM$ and $\psi^t = \varphi^t$, we have the following known results:

THEOREM 5.1 Let $0 < k^2 \leqq -K_x(\pi)$ and let the curvature tensor R have a bounded gradient on M . Then the bundles E^s, E^0, E^u in the Anosov decomposition of $T(SM) = E^s + E^0 + E^u$ are of class C^1 if (i) $\dim M = 2$ (E. Hopf [13]) or if (ii) $0 < k^2 \leqq -K_x(\pi) \leqq K^2$ and $k^2/K^2 > 1/4$ (L. W. Green [6]; cf. also Hirsch and Pugh [12]).

Hopf [13] assumes an upper bound K^2 for $-K_x(\pi)$ in (i), but it will be clear from the proof of Theorem 5.2 (which contains part (i) with $\delta = 1$) that this is not needed. The pinching condition on the curvature (ii) can be written

$$\inf\{K_x(\pi)/K_y(\sigma)\} > 1/4 \quad \text{for } x,y \in M,\ \pi \subset T_x(M),\ \sigma \subset T_y(M). \tag{5.2}$$

Hirsch and Pugh raise the question whether or not this can be relaxed to a uniform "local" pinching condition, say

$$\inf\{K_x(\pi)/K_x(\sigma)\} \geqq \delta^2 \quad \text{for } x \in M \text{ and } \pi,\sigma \in T_x(M), \tag{5.3}$$

where $\delta^2 > 1/4$. We shall answer in part by showing that this is the case if $\delta^2 > 1/2$.

THEOREM 5.2 Let $0 < k^2 \leqq -K_x(\pi) \leqq K^2$ and let the curvature tensor R have a bounded gradient on M . Then E^s, E^0, E^u are of class C^1 if (5.3) holds with $\delta^2 > 1/2$ or, more generally, if (5.3) holds with some $\delta > 0$, $1/4 \leqq \delta^2 < 1$, and the global pinching property (5.2) is relaxed to

$$k^2/K^2 > 1/2 - \delta^2 . \tag{5.4}$$

This reduces to Theorem 5.1 (ii) if $\delta^2 = 1/4$.

PROOF We shall deal only with E^s . In order to use the results of Sections 2-4, it will be convenient to rewrite the Jacobi equation. Let $x = x(t)$ be the geodesic with initial point and initial (unit) tangent vector $x(0)$ and $x'(0)$. Let $\zeta_1(0),\ldots,\zeta_{n+1}(0)$ be an orthonormal basis for $T_{x(0)}(M)$

with $\zeta_{n+1}(0) = x'(0)$, and let $\zeta_1(t),\ldots,\zeta_{n+1}(t)$ be an orthonormal basis for $T_{x(t)}(M)$ obtained by parallel transport, so that $\zeta_{n+1}(t) = T = x'(t)$. If we write

$$(5.5)\qquad z = \sum_{j=1}^{n} v^j(t)\zeta_j(t) + w(t)\zeta_{n+1}(t)\ ,$$

$$(5.6)\qquad D_T z = \sum_{j=1}^{n} v^{j\prime}(t)\zeta_j(t) + w'(t)\zeta_{n+1}(t)\ ,$$

then (5.1) is equivalent to

$$(5.7)\qquad v'' - Q(t)v = 0 \quad \text{and} \quad w'' = 0\ ,$$

where $v = (v^1,\ldots,v^n)$ and $Q(t)$ is the symmetric matrix

$$(5.8)\qquad Q(t) = -(\langle R(\zeta_i,T)T,\zeta_j\rangle) \quad \text{for } i,j = 1,\ldots,n\ ,$$

and $\langle\cdot,\cdot\rangle$ is the scalar product on $T(M)$; cf., e.g., [4], p. 128. If $|\xi| = 1$, then $-(Q(t)\xi,\xi) = \langle R(\zeta,T)T,\zeta\rangle = K_{x(t)}(\pi)$, where π is the 2-plane in $T_{x(t)}(M)$ determined by the orthonormal vectors T and $\zeta = \Sigma\xi^j\zeta_j(t)$, so that $Q(t) \geqq k^2 I > 0$.

We shall use the notation of Proposition 4.3. Since $\langle z,z\rangle = |v|^2 + w^2$, the only solutions of (5.1) which are exponentially small at $t = \infty$ correspond to $w \equiv 0$ and $v = Y(t)\xi$, so that $v' = Y'(t)\xi = Y'Y^{-1}v$. This gives us the existence of the Anosov decomposition $T(SM) = E^s + E^0 + E^u$, with E^s having the n-dimensional fibre $E^s_{(x(0),x'(0))} = \{(z,z^1) : \langle z,x'(0)\rangle = 0\ ,\ z^1 = A(x(0),x'(0))z\}$, where $A(x(0),x'(0))$ is a linear (symmetric) nonsingular mapping of the vector space $T^{\perp}_{x(0)}(M) = \{x' \in T_{x(0)}(M) : \langle x',x'(0)\rangle = 0\}$ which in terms of the basis $\zeta_1(0),\ldots,\zeta_n(0)$ for $T^{\perp}_{x(0)}(M)$ is given by the matrix $A = Y'(0)Y^{-1}(0)$. The uniqueness of the principal solution Y (up to constant nonsingular factors on the right) makes it clear that $E^s_{(x(0),x'(0))}$ is continuous on SM .

The bundle E^s is of class C^1 if and only if $A(x(0),x'(0)) = Y'(0)Y^{-1}(0)$ is of class C^1 as a function of $(x(0),x'(0)) \in SM$. This is the case if, for every smooth family of smooth arcs $(x(0,\alpha),x'(0,\alpha)) \in SM$ for small $\alpha \geqq 0$, $A(x(0,\alpha),x'(0,\alpha))$ has a derivative with respect to α (say, at $\alpha = 0$) which is a continuous function of $(x(0),x'(0)) = (x(0,0),x'(0,0)) \in SM$. Actually, it suffices to consider only two types of arcs; (a) $(x(0,\alpha),x'(0,\alpha)) = (x(0),x'(0,\alpha))$ and $x'(0,\alpha)$ is the rotation of $x'(0)$ through an angle α in a fixed 2-plane in $T_{x(0)}(M)$ and (b) $x(0,\alpha)$ is a unit speed geodesic, $\langle \partial x(0,\alpha)/\partial\alpha, x'(0)\rangle = 0$ at $\alpha = 0$, and $x'(0,\alpha)$ is obtained by parallel transport of $x'(0) = x'(0,0)$. Correspondingly, let $\zeta_1(0,\alpha),\ldots,\zeta_{n+1}(0,\alpha)$ be a C^1 orthonormal basis of $T_{x(0,\alpha)}(M)$ with $\zeta_{n+1}(0,\alpha) = T = x'(0,\alpha)$ and let $\zeta_1(t,\alpha),\ldots,\zeta_{n+1}(t,\alpha)$ be obtained by parallel transport along the geodesic $x = x(\cdot,\alpha)$ with initial point $(x(0,\alpha),x'(0,\alpha))$. The Jacobi equation along this geodesic takes the form (5.7) - (5.8), where $\zeta_j(t) = \zeta_j(t,\alpha)$ and $Q(t) = Q(t,\alpha)$.

In order to prove Theorem 5.2, it suffices to show that if $Y(t) = Y(t,\alpha)$ is the principal solution given by Proposition 4.3, then Y,Y' are of class C^1 and that $Y(0,0),Y'(0,0)$ are C^1 functions of $(x(0),x'(0)) \in SM$. This will be the case if the estimates in the assumptions of Corollary 4.1 are valid uniformly in α and $(x(0),x'(0))$. Actually, we have $|X(t)| \cdot |X^{-1}(t)|^{1/\theta} \leqq 1$ with $\theta > 1/2$ trivially if $n = 1$ or by Theorem 3.1 if $n > 1$, since (5.3) implies the condition $p/q \geqq \delta^2$ and (5.4) implies (2.5) with $\theta > 1/2$ (and k,K,δ,θ independent of $\alpha,x(0),x'(0)$). It only remains to verify

$|Q_\alpha| \leq C|X|$ with C independent of $\alpha, x(0), x'(0)$.

If $\dim M = 2$, Q is a scalar $Q = -K(x(t,\alpha))$, where $K(x)$ is the Gauss curvature at $x \in M$. In this case, $Q_\alpha = -\nabla K \cdot \partial x/\partial\alpha$. By assumption, the gradient ∇K is bounded on M . Also $z = \partial x/\partial\alpha$ is a solution of the Jacobi equation (5.1). From the choices (a) and (b) of the curves $(x(0,\alpha), x'(0,\alpha))$ above, it is clear that $\langle z, x'(t,\alpha)\rangle = 0$, i.e., $w(t) \equiv 0$ in (5.5), so that z is of the form $z = X(t)\xi + Y(t)\eta$. Since $X'Y - Y'X = 1$ and $0 < Y \leq X$,it is clear that there exists a suitable C such that $|Q_\alpha| \leq C|X|$.

If $\dim M > 2$, let $W = \partial/\partial\alpha$. If we differentiate the (i,j)-th element $\langle R(\zeta_j, T)T, \zeta_i\rangle$ of $-Q(t)$, we get the sum of terms resulting from the application of D_W to $R, \zeta_j, T, T, \zeta_i$. As in the case $n = 2$, the term involving $D_W R$ has a majorant of the form $C|X|$. Since $D_W T$ corresponds to $(\partial x(t,\alpha)/\partial\alpha)'$, which is a vector of the form $X'\xi + Y'\eta$, it follows from Corollary 3.3 and Propositions 4.1 - 4.2, that $|D_W T| \leq C|X|$. To estimate $D_W\zeta_j$, note that $D_T D_W \zeta_j = R(T,W)\zeta_j$ since $D_T\zeta_j = 0$. We have $|W| \leq C|t|$ or $|W| = 1$ in the choices (a) or (b) of curves $(x(0,\alpha), x'(0,\alpha))$. Thus $|D_T D_W \zeta_j| \leq C(1 + t)$ and $|D_W\zeta_j| \leq C(1 + t^2) \leq C|X|$ for suitable constants C . This completes the proof since $k^2 \leq -K_x(\pi) \leq K^2$ implies that R is bounded on M .

REFERENCES

0. Ahlbrandt, C. D. Linear independence of the principal solutions at ∞ and $-\infty$ for formally self-adjoint differential systems, to appear.

1. Anosov, D. Geodesic flows on closed Riemann manifolds with negative curvature, *Proc. Steklov Inst. Math.* *90*(1967): English translation, *Amer. Math. Soc.*, 1969.

2. Anosov D. and Ya. Sinai. Some smooth ergodic systems, *Uspekhi Mat. Nauk* *22*(1967), pp. 107-172: *Russian Math. Surveys* *22*(1967), pp. 103-167.

3. Berger, M. An extension of Rauch's metric comparison theorem and some applications, *Illinois J. Math.* *6*(1962), pp. 700-712.

4. Coppel, W. A. "Disconjugacy." Lecture Notes in Mathematics No. 220, Springer, 1971.

5. Green, L. W. A theorem of E. Hopf, *Michigan Math. J.* *5*(1958), pp. 31-34.

6. Green, L. W. The generalized geodesic flow, *Duke Math. J.* *41*(1974), pp. 115-126; *42*(1975), p. 381.

7. Gromoll, D., W. Klingenberg and W. Meyer. "Riemannische Geometrie im Grossen." 2nd ed. Lecture Notes in Mathematics No. 55, Springer, 1975.

8. Hartman, P. Unrestricted solution fields of almost separable equations, *Trans. Amer. Math. Soc.* *63*(1948), pp. 560-580.

9. Hartman, P. Self-adjoint, non-oscillatory systems of ordinary, second order, linear differential equations, *Duke Math. J.* *24*(1957), pp. 25-36.

10. Hartman, P. "Ordinary Differential Equations." S. M. Hartman, Baltimore, 1973.

11. Hartman, P. and A. Wintner. On nonoscillatory linear differential equations with monotone coefficients, *Amer. J. Math.* *76*(1954), pp. 207-219.

12. Hirsch M. W. and C. C. Pugh. Smoothness of horocycle foliations, *J. Differential Geometry* *10*(1975), pp. 225-238.

13. Hopf, E. Statistik der Lösungen geodätischer Probleme vom unstabilen Typus II, *Math. Ann.* *117*(1940), pp. 590-608.

14. Hopf. E. Closed surfaces without conjugate points, *Proc. Nat. Acad. Sci. U.S.A.* *34*(1948), pp. 47-51.

15. Morse, M. A generalization of the Sturm separation and comparison theorems in n-space, *Math. Ann.* *103*(1930), pp. 52-59.

16. Rauch, H. E. A contribution to Riemannian geometry in the large, *Ann. Math.* *54*(1951), pp. 38-55.

17. Reid, W. T. "Ordinary Differential Equations." Wiley, New York, 1971.

COMPARISON CRITERIA FOR DISCONJUGACY

James S. Muldowney

The University of Alberta
Edmonton, Alberta, Canada

Let I denote a real interval and L the n-th order real linear differential operator defined by

(1) $$Ly = y^{(n)} + a_1(t)y^{(n-1)} + \cdots + a_n(t)y$$

where the coefficients a_i are required to be continuous on I.

DEFINITION L is *disconjugate* on I if the only solution of $Ly = 0$ having n zeros or more in I counting multiplicities is the zero solution.

A survey of results on disconjugacy for L may be found in the book of Coppel [3]. Most disconjugacy criteria involve a comparison of L with another operator for which the behaviour of the null set is known. For example, it is clear that the operator D^n is disconjugate on any real interval and many disconjugacy conditions in the literature (cf. [4],[6],[21]) require that the coefficients $a_i(t)$ in (1) be small in some

ISBN 0-12-632250-3

sense, i.e., L is regarded as a small perturbation of D^n. The Sturm Theorem gives a general comparison criterion for second order operators. For $n > 2$ a theorem of this type was first formulated by Levin [10] without a proof; the first proof was given by Nehari [14]. The Levin-Nehari Theorem states that if $q_{n,1}(t) \leq 0 \leq q_{n,2}(t)$, $t \in I$, $n > 2$, then the disconjugacy of $L + q_{n,i}$ on $[a,b]$, $i = 1,2$, implies the disconjugacy of L on $[a,b]$. More recently comparison theorems of a more general type for higher order operators were formulated and proved independently by Levin [11] and Hartman [7], [8]. These results are generalizations of the following form of Sturm's Theorem: If $n = 2$ then L is disconjugate on $[a,b]$ if and only if there exists $u \in C^2[a,b)$ such that

$$u > 0 , \quad Lu \leq 0 \quad \text{on} \quad [a,b) .$$

A good exposition of the work of Hartman and Levin may be found in Coppel's book [3]. The criterion stated in Theorem 1 may be found in Hartman [8]. Here $W(u_1,\ldots,u_n)$ denotes the Wronskian determinant of the functions $(u_1,\ldots,u_n)$ and the symbol $(u_1,\ldots,\hat{u}_j,\ldots,u_n)$ denotes the $(n - 1)$-tuple $(u_1,\ldots,u_{j-1},u_{j+1},\ldots,u_n)$.

THEOREM 1 L is disconjugate on $[a,b]$ if and only if there exist $u_1,\ldots,u_{n-1} \in C^n[a,b)$ such that

(i) $W(u_1,\ldots,u_k) > 0$ on $[a,b)$, $k = 1,\ldots,n - 1$

(ii) $W(u_1,\ldots,\hat{u}_j,\ldots,u_k)(a) \geq 0$, $1 \leq j < k < n$

(iii) $(-1)^{n-k} Lu_k \geq 0$, on $[a,b)$.

For example, if one chooses $u_i(t) = e^{\alpha_i t}$, $i = 1,\ldots,n-1$, where $\alpha_1 < \cdots < \alpha_{n-1}$ it follows that L is disconjugate on $[a,b]$ if $\lambda_1(t) \leq \alpha \leq \lambda_2(t) \leq \cdots \leq \lambda_{n-1}(t) \leq \alpha_{n-1} \leq \lambda_n(t)$

here $\lambda = \lambda_i(t)$ are the zeros of the polynomial $(\lambda,t) = e^{-\lambda t}L(e^{\lambda t})$. Any operator M of order n which conains the elements $(u_1,\ldots,u_{n-1})$ in its null set is disconjuate on $[a,b]$ and Theorem 1 gives a comparison of L with any uch operator.

A comparison theorem which uses n test functions u_i ather than n - 1 , as in Theorem 1, may be deduced from nonscillation conditions developed by Nehari [15], [16], Schwarz 19] and Friedland [5] for first order linear systems of diferential equations. In these papers best possible constants $_{rp}(n)$ are found such that

$$\int_a^b \|A\|_{rp} < c_{rp}(n)$$

implies that the only solution $x = \mathrm{col}(x_1,\ldots,x_n)$ of $x' = A(t)x$ which satisfies $x_i(t_i) = 0$, $i = 1,\ldots,n$ for any $\{t_1,\ldots,t_n\} \subset [a,b]$ is the zero solution. Here $\|A\|_{rp} = \sup\|Ax\|_p/\|x\|_r$, $1 \le p \le \infty$, $1 \le r \le \infty$ and $\|\cdot\|_s$ denotes the usual Hölder norm in $\mathbb{R}^n$. In particular, it is shown that if $1 \le p \le r \le \infty$, then

$$c_{rp}(n) = 2\int_0^1 (1+\tau^r)^{-(1/r)}(1+\tau^q)^{-(1/q)}d\tau\ , \quad p^{-1}+q^{-1}=1\ ,$$

$$c_{r\infty}(2m) = 2m^{-(1/r)}\int_0^1 [(1-\tau^r)^{1/r}(1+\tau)]^{-1}d\tau\ ,$$

$$c_{r\infty}(2m+1) = m^{-(1/r)}\int_0^1 \left[\left(\tau^r + \frac{m+1}{m}\right)^{-(1/r)} + \left(\frac{m+1}{m}\tau^r + 1\right)^{-(1/r)}\right](1+\tau)^{-1}d\tau\ .$$

The following theorem is proved in [12].

THEOREM 2 Suppose $u_1,\dots,u_n \in C^n[a,b]$ are such that

(i) $W(u_1,\dots,u_k) > 0$ on $[a,b]$, $1 \le k \le n$,

(ii) $W(u_1,\dots,\hat{u}_j,\dots,u_k) > 0$ on $[a,b]$, $1 \le j < k \le n$,

and $u = (u_1,\dots,u_n)$, $v = (v_1,\dots,v_n)$ where

$$v_j = W(u_1,\dots,\hat{u}_j,\dots,u_n)/W(u_1,\dots,u_n) .$$

Then L is disconjugate on $[a,b]$ if

$$\int_a^b \|v\|_p \|Lu\|_s < c_{rp}(n) , \quad r^{-1} + s^{-1} = 1 . \tag{2}$$

This result is proved by showing, with the aid of the Pólya Mean Value Theorem [18], that if y is a nontrivial solution of $Ly = 0$ having n zeros or more in $[a,b]$ and

$$x_j = W(u_1,\dots,\hat{u}_j,\dots,u_n,y)/W(u_1,\dots,u_n) , \quad j = 1,\dots,n ,$$

then each x_j vanishes at some point in $[a,b]$ and, in addition, satisfies the equation

$$x_j' = v_j \sum_{k=1}^{n} (-1)^{n-k-1}(Lu_k)x_k , \quad j = 1,\dots,n .$$

Thus, any nonoscillation condition (such as (2)) for this system is a disconjugacy criterion for L. For example, if one chooses $r = p = 2$ and $u_i(t) = e^{\alpha_i t}$, $i = 1,\dots,n$, ($\alpha_i \ne \alpha_j$, $i \ne j$) then, since $c_{22}(n) = \frac{\pi}{2}$,

$$\int_a^b \left[\sum_{k=1}^{n} \beta_k^2 e^{-2\alpha_k t} \sum_{j=1}^{n} P(\alpha_j,t)^2 e^{2\alpha_j t}\right]^{\frac{1}{2}} dt < \frac{\pi}{2}$$

implies L is disconjugate on $[a,b]$, where

$$P(\lambda,t) = e^{-\lambda t}L(e^{\lambda t}) \quad \text{and} \quad \beta_k = \prod_{j\ne k} (\alpha_j - \alpha_k)^{-1} .$$

A result of Levin [10], with first published proof due to Sherman [20], plays an important role in the theory of disconjugacy. This result states that L is disconjugate on I if and only if, for each $[a,b] \subset I$ and $k = 1,\ldots,n-1$ there is no function y which is positive on (a,b) and satisfies

$$(3) \qquad Ly = 0 \,, \quad y(a) = \cdots = y^{(k-1)}(a) = y(b) = \cdots = y^{(n-k-1)}(b) = 0 \,.$$

In the remaining remarks it is assumed that the coefficients $a_k(t)$ are sufficiently smooth that $C^n[a,b]$ is in the domain of the formal adjoint L^* of L.

LEMMA If there exists $\psi \in C^n[a,b]$ such that

$$(4) \qquad L^*\psi \geq 0 \; (\not\equiv 0) \,, \quad \psi(a) = \cdots = \psi^{n-k-1}(a) = \psi(b) = \cdots = \psi^{k-1}(b) = 0 \,,$$

then there is no solution of (3) which is positive on (a,b).

This follows from the contradiction that if y were such a solution then, from Lagrange's Identity,

$$\begin{aligned} 0 &= [y\psi](b) - [y\psi](a) \\ &= \int_a^b (\psi Ly - yL^*\psi) \\ &= -\int_a^b yL^*\psi < 0 \,. \end{aligned}$$

Here $[y\psi] = \sum_{m=1}^{n} \sum_{j+k=m-1} (-1)^j y^{(k)}(a_{n-m}\psi)^{(j)}$, $a_0 = 1$.

From this lemma and the theorem of Levin and Sherman the following theorem is deduced.

THEOREM 3 A sufficient condition for the disconjugacy of L on I is that for each $[a,b] \subset I$ and $k = 1,\dots,n-1$ there exists $\psi \in C^n[a,b]$ satisfying (4). A necessary condition is that such functions should exist and satisfy $(-1)^{n-k}\psi > 0$ on (a,b).

One can deduce from Theorem 3, for example, the Levin-Nehari Comparison Theorem mentioned earlier. If $q_{n,1}(t) \le 0$ and $L + q_{n,1}$ is disconjugate on I so also is $L^* + q_{n,1}$ and, for $[a,b] \subset I$, the solution of

$$(L^* + q_{n,1})\psi = 1, \quad \psi(a) = \cdots = \psi^{n-k-1}(a) = \psi(b) = \cdots = \psi^{(k-1)}(b) = 0$$

satisfies $(-1)^{n-k}\psi > 0$ on (a,b) so that, if $n-k$ is even, $L^*\psi = 1 - q_{n,1}\psi > 0$ and (4) is satisfied in this case. Similarly if $q_{n,2}(t) \ge 0$ one can find, when $n-k$ is odd, functions ψ satisfying (4).

A result of Bessmertnyh and Levin [2] states that if

$$\sum_{j=1}^{n-1} \frac{n-j}{j!\,n} |a_j(t)|(b-a)^j \tag{5}$$
$$+ \frac{(n-1)^{n-1}}{n!\,n^n} |a_n(t)|(b-a)^n \le 1$$

then L is disconjugate on $[a,b]$. This follows easily from Theorem 3 by considering the functions

$$\psi_k(t) = (t-a')^k(b'-t)^{n-k}/n!, \quad k = 1,\dots,n-1$$

for each $[a',b'] \subset [a,b]$ and by showing that the coefficient of $|a_j(t)|$ in (5) is an upper bound for $|\psi_k^{(n-j)}(t)|$, $t \in [a',b']$.

Theorem 4, which is proved in [13], may be considered a omparison result in the spirit of Liapunov's Inequality. It is hown that if $H(t,s)$ is a function of a certain type, which ncludes all Green's functions for $(k,n-k)$ interpolation roblems on $[a,b]$, then the boundary value problem (3) has no olution y such that $y > 0$ on (a,b) provided $L_s^* H(t,s)]_-$ is 'small' in some sense on $[a,b] \times [a,b]$. urther it is shown that if $|L_s^* H(t,s)|$ is 'small' then (3) has o nontrivial solution.

Let $\mathcal{H}(n-k;a,b)$ denote the set of functions $H : [a,b] \times$ a,b] $\to \mathbb{R}$ such that, for each $t \in [a,b]$,

(i) $H(t,\cdot) \in C^n([a,t) \cup (t,b]) \cap C^{n-2}[a,b]$

(ii) $$\frac{\partial^{n-1}}{\partial s^{n-1}} H(t,t+) - \frac{\partial^{n-1}}{\partial s^{n-1}} H(t,t-) = (-1)^n$$

(iii) $$H(t,a) = \cdots = \frac{\partial^{n-k-1}}{\partial s^{n-k-1}} H(t,a)$$

$$= H(t,b) = \cdots = \frac{\partial^{k-1}}{\partial s^{k-1}} H(t,b) = 0 \quad .$$

f $x \in C^n[a,b]$ and $x(a) = \cdots = x^{(k-1)}(a) = x(b) = \cdots =$ $^{(n-k-1)}(b) = 0$, then from Lagrange's Identity

$$\int_a^b H(t,s)Lx(s)ds - \int_a^b x(s)L_s^* H(t,s)ds$$

$$= [xH(t,\cdot)](b) - [xH(t,\cdot)](t+)$$

$$+ [xH(t,\cdot)](t-) - [xH(t,\cdot)](a)$$

6) $$= x(t) \ , \quad a \le t \le b$$

ince $[xH(t,\cdot)](b) - [xH(t,\cdot)](a) = 0$ and $[xH(t,\cdot)](t-) -$ $[xH(t,\cdot)](t+) = x(t)$. Thus any solution of $Ly = 0$ having k

zeros at a and n - k zeros at b satisfies

(7) $$y(t) = -\int_a^b y(s)L_s^* H(t,s)ds \ , \quad a \le t \le b \ .$$

THEOREM 4 Suppose $H \in \mathcal{H}(n-k;a,b)$ and $\eta : [a,b] \to \mathbb{R}$ satisfies $\eta \ge 0$ and $\eta y \in L_p[a,b]$ if $y \in C^n[a,b]$ and y has k zeros at a , n - k zeros at b . Then

(8) $$\int_a^b \{\int_a^b [L_s^* H(t,s)\eta(t)/\eta(s)]_-^q ds\}^{p/q} dt < 1$$

$1 < p < \infty$, $p^{-1} + q^{-1} = 1$, implies that the problem (3) has no solution which is positive on (a,b) . Moreover

(9) $$\int_a^b \{\int_a^b |L_s^* H(t,s)\eta(t)/\eta(s)|^q ds\}^{p/q} dt < 1$$

implies (3) has no nontrivial solution. If p or $q = \infty$ the corresponding integral in (8), (9) should be replaced by the appropriate supremum.

This follows from (7) by observing that if y satisfies (3) then $x = \eta y$ is a fixed point of the map $T : L_p[a,b] \to L_p[a,b]$ defined by

$$(Tz)(t) = -\int_a^b z(s)L_s^* H(t,s)\eta(t)/\eta(s)ds \ , \quad a \le t \le b \ .$$

Condition (8) implies that 0 is the only nonnegative fixed point and (9) implies that 0 is the only fixed point of T .

If one considers $H(t,s) = G(t,s) - \psi(s)$ where

$$-G(t,s)(b-a) = \begin{cases} (b-t)(s-a), & a \le s \le t \\ (b-s)(t-a), & t \le s \le b \end{cases}$$

$$\psi \in C^2[a,b], \quad \psi(a) = \psi(b) = 0$$

and $Ly = L^*y = y'' + q(t)y$, then $H \in \mathcal{H}(1;a,b)$ and one finds that the boundary value problem

$$Ly = 0, \quad y(a) = y(b) = 0$$

has no solution such that $y > 0$ on (a,b) if

$$\int_a^b \left[q_+(s)\,\frac{(b-s)(s-a)}{b-a} + \psi''(s) + q(s)\psi(s)\right]_+ ds \le 1 .$$

The case $\psi = 0$ is Liapunov's Inequality [cf. 9, Theorem 5.1, p. 345]. It should be noted however that unlike Liapunov's Inequality this condition does not imply the disconjugacy of L on $[a,b]$ since the existence of ψ on $[a,b]$ does not imply the existence of a similar function for each subinterval of $[a,b]$.

Theorem 4 may also be used to establish the Levin-Nehari Comparison Principle and has the added advantage of proving a monotone dependence of Green's functions on their corresponding differential operators as is shown in the following corollary.

COROLLARY Suppose $H \in \mathcal{H}(n-k;a,b)$ is a Green's function satisfying

$$(-1)^{n-k}H(t,s) \ge 0, \quad L_s^*H(t,s) \le 0$$

(10) $$\text{on } [a,b] \times [a,b] .$$

Suppose further that there exists a function φ such that

$$L\varphi \geq 0 \; (\not\equiv 0) \, , \quad \varphi(a) = \cdots = \varphi^{(k-1)}(a) = \varphi(b)$$

$$= \cdots = \varphi^{(n-k-1)}(b) = 0 \tag{11}$$

with $\varphi^{(k)}(a) \neq 0$, $\varphi^{(n-k)}(b) \neq 0$, $(-1)^{n-k}\varphi > 0$ on (a,b). Then Green's function for the problem

$$Ly = f \, , \quad y(a) = \cdots = y^{(k-1)}(a)$$

$$= y(b) = \cdots = y^{(n-k-1)}(b) = 0 \tag{12}$$

exists and satisfies

$$(-1)^{n-k}G(t,s) \geq (-1)^{n-k}H(t,s) \, .$$

PROOF From (11) and (6) it follows that

$$\varphi(t) = \int_a^b H(t,s)L\varphi(s)ds - \int_a^b \varphi(s)L_s^* H(t,s)ds$$

and so, from (9),

$$\int_a^b \frac{\varphi(s)}{\varphi(t)} \, |L_s^* H(t,s)| ds = 1 - \frac{1}{\varphi(t)} \int_a^b H(t,s)L\varphi(s)ds \, .$$

The right-hand side is a continuous function of t and has supremum less than 1 and so (9) holds with $\eta = \frac{1}{\varphi}$, $p = \infty$. Thus Green's function $G(t,s)$ for the problem (12) exists. In the same way that (6) was proved it may be shown that

$$G(t,\tau) = H(t,\tau) - \int_a^b G(s,\tau)L_s^* H(t,s)ds$$

and solving this equation for G by Picard iterations shows that

$$(-1)^{n-k}G(t,\tau) \geq (-1)^{n-k}H(t,\tau) \geq 0 \, .$$

In order to increase the utility of Theorem 4 it is desirable that a comprehensive system of inequalities be obtained for the quantities $\left| \frac{\partial^j}{\partial s^j} H(t,s) \right|$, $j = 0,\ldots,n$, when H is any Green's function. Such inequalities were obtained by Ostroumov [17] for Green's functions corresponding to the operator D^n and similar estimates were obtained by Bates and Gustafson [1] in the case $j = 0$ for more general Green's functions.

REFERENCES

1. Bates, P. W. and G. B. Gustafson. Green's function inequalities for two-point boundary value problems, *Pacific J. Math. 59*(1975), pp. 327-343.

2. Bessmertnyh, G. A. and A. Ju. Levin. Some inequalities satisfied by differentiable functions of one variable, *Dokl. Akad. Nauk SSSR 144*(1962), pp. 471-474: *Soviet Math. Dokl. 3*(1962), pp. 737-740.

3. Coppel, W. A. "Disconjugacy." Springer-Verlag, New York, 1971.

4. Fink, A. M. Differential inequalities and disconjugacy, *J. Math. Anal. Appl. 49*(1975), pp. 758-772.

5. Friedland, S. Nonoscillation and integral inequalities, *Bull. Amer. Math. Soc. 80*(1974), pp. 715-717.

6. Hartman, P. On disconjugacy criteria, *Proc. Amer. Math. Soc. 24*(1970), pp. 374-381.

7. Hartman, P. Principal solutions of disconjugate n-th order linear differential equations, *Amer. J. Math. 91*(1969), pp. 306-362: Corrigendum and addendum, ibid. 93(1971), pp. 438-451.

8. Hartman, P. Disconjugacy and Wronskians, "Japan-United States Seminar on Ordinary Differential and Functional Equations," (M. Urabe, editor), pp. 208-218. Springer-Verlag, New York, 1971.

9. Hartman, P. "Ordinary Differential Equations." Hartman, Baltimore, 1973.

10. Levin, A. Ju. Some problems bearing on the oscillation of solutions of linear differential equations, *Dokl. Akad. Nauk SSSR 148*(1963), pp. 512-515: *Soviet Math. Dokl. 4*(1963), pp. 121-124.

11. Levin, A. Ju. Non-oscillation of solutions of the equation $x^{(n)} + p_1(t)x^{(n-1)} + \cdots + p_n(t)x = 0$, *Uspehi Mat. Nauk 24*(1969), pp. 43-96: *Russian Math. Surveys 24*(1969), pp. 43-99.

12. Muldowney, J. S. A disconjugacy criterion for linear differential operators, to appear *Proc. Amer. Math. Soc.*

13. Muldowney, J. S. Comparison Theorems for linear boundary value problems, to appear *SIAM J. Math. Anal.*

14. Nehari, Z. Disconjugate linear differential operators, *Trans. Amer. Math. Soc. 129*(1967), pp. 500-576.

15. Nehari, Z. Oscillation theorems for systems of linear differential equations, *Trans. Amer. Math. Soc. 139*(1969), pp. 339-347.

16. Nehari, Z. Nonoscillation and disconjugacy of systems of linear differential equations, *J. Math. Anal. Appl. 42* (1973), pp. 237-254.

17. Ostroumov, V. V. Unique solvability of the de la Vallee Poussin problem, *Differentsial'nye Uravneniya 4*(1968), pp. 261-268: *Differential Equations 4*(1968), pp. 135-139.

18. Pólya, G. On the mean-value theorem corresponding to a given linear differential equation, *Trans. Amer. Math. Soc. 24*(1924), pp. 312-324.

19. Schwarz, B. Norm conditions for disconjugacy of complex differential systems, *J. Math. Anal. Appl. 28*(1969), pp. 553-568.

20. Sherman, T. L. Properties of solutions of n-th order linear differential equations, *Pacific J. Math.* *15*(1965), pp. 1045-1060.

21. Willett, D. Generalized de la Vallee Poussin disconjugacy tests for linear differential equations, *Canad. Math. Bull.* *14*(1971), pp. 419-428.

OPTIMAL CONTROL AND DIFFERENTIAL EQUATIONS

POSITIVELY INVARIANT SETS FOR DIFFERENTIAL EQUATIONS

George Seifert

Iowa State University
Ames, Iowa

Suppose a biological process is modelled by a system of differential equations

$$x_i' = f_i(t,x_1,\dots,x_n)\ , \quad i = 1,2,\dots,n \tag{1}$$

where the prime here and henceforth denotes differentiation with respect to t and the x_j represent populations of certain distinct interacting species varying with time t. Such a mathematical model is clearly valid only if any solution $(x_1(t),\dots,x_n(t))$ with initial values $x_j(t_0) = x_{j0} \geq 0$ at $t = t_0$ satisfies $x_j(t) \geq 0$, $j = 1,\dots,n$ for all $t > t_0$ in some interval about t_0. Thus conditions necessary and sufficient for such behavior of solutions of (1) may be of interest.

More generally, let us consider the ordinary differential equation

$$x' = f(t,x) \tag{2}$$

ISBN 0-12-632250-3

where x and $f(t,x)$ have values in some Banach space X. Let $M \subset X$; we define M to be positively invariant for (2) if for any real t_0 and every solution $x(t)$ of (2) with $x(t_0) \in M$, we have $x(t) \in M$ for $t > t_0$ as long as $x(t)$ exists. Clearly if M is open, it will be positively invariant for (2) if and only if the Cauchy initial value problem for (2) has a solution. So we henceforth suppose M to be closed.

If $X = R^n$, real Euclidean n-space, and if f is continuous on $R \times M$, a necessary and sufficient condition for the positive invariance of M was given by Nagumo in 1942 [1]. More recently, others have published conditions of essentially the same or equivalent type for the case $X = R^n$; cf. Yorke [2], Brezis [3], Bony [4], Crandall [5], Hartman [6]. For X not necessarily finite dimensional, similar but stronger conditions seem to be required; cf. Martin [7], Volkman [8]. In case X is a Hilbert space, inner product conditions such as used by Bony [4] for $X = R^n$ have also been used for X infinite dimensional, but involve stronger conditions on M. For a good discussion of these results in R^n, cf. Redheffer [9].

Suppose now that the solutions of (2) are uniquely determined by their initial values. Then Nagumo's condition necessary and sufficient that M be positively invariant for (2) is essentially as follows: For each $\varepsilon > 0$ and each $(t_0,x_0) \in R^1 \times M$, there exists $(t_1,x_1) \in R^1 \times M$, $t_0 < t_1 < t_0 + \varepsilon$ such that

$$\left| \frac{x_1 - x_0}{t_1 - t_0} - f(t_0,x_0) \right| < \varepsilon . \tag{3}$$

The so-called subtangential condition used by Brezis

$$\lim_{h \to 0+} h^{-1} \operatorname{dist}(x + hf(t,x), M) = 0 \tag{4}$$

or each $(t,x) \in R \times M$, is equivalent to (3).

If y is a point on the boundary ∂M of M such that here exists a $x_0 \notin M$ for which $|x - x_0| < |y - x_0|$ implies $\notin M$, then the condition

$$\langle f(t,y), x_0 - y \rangle \le 0 \tag{5}$$

t all such points $y \in \partial M$ is also necessary and sufficient for he positive invariance of M under the additional condition hat f be locally Lipschitzian in x ; here $\langle x,y \rangle$ denotes he inner product of x and y . This inner product condition, sed by Bony [4], also can be shown to be equivalent to Nagumo's ondition (3) under the above condition on f .

Suppose we now consider an equation like (2) but with a time delay; i.e., a functional differential equation of retarded type:

$$x'(t) = f(t,x_t) \tag{6}$$

where x_t denotes the function $x_t(s) = x(t + s)$, $s \le 0$, $x(t)$ a function on R^1 to X ; now f is defined on $F \times R$, where F is some suitable space of functions on some subset of the half-line $t \le 0$. A fairly large class of such functions which can be made into a Banach space is the set of functions φ continuous and bounded on $t \le 0$; we henceforth denote this class by CB and use it for F . Suppose M is closed, $M \subset X$, and $\varphi \in CB$ also satisfies $\varphi(s) \in M$ for $s \le 0$. Then if for each t_0 and solution $x(t)$ of (6) such that $x(t_0 + s) = \varphi(s)$, $s \le 0$, we have $x(t) \in M$ for all $t > t_0$ for which it is defined, we say M is positively invariant for (6). The question of conditions for the positive invariance of M for (6) have apparently only recently been studied even for the case $X = R^n$. However, the natural adaption of Nagumo's condition to the delay case (3) does not produce such a condition for arbitrary sets M .

The natural adaptation of (4) to the delay case (6) is:

(7) $$\lim_{h\to 0+} h^{-1}\operatorname{dist}(\varphi(0) + hf(t,\varphi),M) = 0$$

for all $(t,\varphi) \in R^1 \times CB$ with $\varphi(s) \in M$ for $s \le 0$.

The following simple example shows that (7) does not imply the positive invariance of M for (6). Take $X = R^1$, and $M = \{x \le 0\} \cup \{n^{-1} : n = 1,2,\ldots\}$. Define

$$f(t,\varphi) = -\varphi(-1) \quad \text{if} \quad \varphi(-1) < 0 \quad \text{and}$$
$$f(t,\varphi) = 0 \quad \text{if} \quad \varphi(-1) \ge 0 .$$

Let $\varphi(s) \in M$ for $s \le 0$. Then either $\varphi(s) \le 0$ for $s \le 0$ or there exists an integer $n \ge 1$ such that $\varphi(s) = n^{-1}$ for $s \le 0$. In the second case (7) holds trivially. In the first case if $\varphi(0) < 0$, (7) also holds trivially; if $\varphi(0) = 0$ it can be shown that

$$h^{-1}\inf\{|-h\varphi(-1) - n^{-1}| : n = 1,2,\ldots\} \to 0 \quad \text{as} \quad h \to 0+ ;$$

so (7) holds in any case. But M is not positively invariant for (6); if $\varphi \in CB$ is such that $\varphi(s) \in M$, $\varphi(0) = 0$, and $\varphi(-1) < 0$, then if $x(t)$ is a solution of (6) for $t \ge 0$ with $x(s) = \varphi(s)$, $s \le 0$, $x'(0)$ must exist at least as a right hand derivative, and $x'(0) > 0$. So $x(t)$ must leave M as t increases from 0 .

Examples of connected sets M and delay equations in R^2 exist which also show that (7) fails to be a condition sufficient for the positive invariance of M for such equations even if M is connected.

In a recent paper [10], the author has shown that for convex sets M and suitably restricted functions f , (7) does turn out to be necessary and sufficient for the positive invariance of M for (6). Applications to Volterra integro-differential equations are given in [10]. Other as yet unpublished results, in terms of inner product conditions, are also known even for the case where X is an infinite dimensional Hilbert space.

REFERENCES

1. Nagumo, M. Über die Lage der Integralkurven gewöhnlicher Differentialgleichungen, *Proc. Physico - Math. Soc. Japan, ser. 3,24*(1942), pp. 550-559.

2. Yorke, J. A. Differential inequalities and non-Lipschitz scalar functions, *Math. Systems Theory 4*(1970), pp. 140-153.

3. Brezis, H. On a characterization of flow invariant sets, *Comm. Pure Appl. Math. 23*(1970), pp. 261-263.

4. Bony, J.-M. Principe du maximum, inequalite de Harnack et unicite du problemes de Cauchy pour les operateurs elliptiques degeneres, *Ann. Inst. Fourier Grenoble 19*(1969), pp. 227-304.

5. Crandall, M. G. A generalization of Peano's existence theorem and flow invariance, *Proc. Amer. Math. Soc. 36* (1972), pp. 151-155.

6. Hartman, P. On invariant sets and on a theorem of Wazewski, *Proc. Amer. Math. Soc. 32*(1972), pp. 511-520.

7. Martin, R. H. Approximation and existence of solutions to ordinary differential equations in Banach spaces, *Funk. Ekv. 16*(1973), pp. 195-211.

8. Volkmann, P. Über die positive Invarianz einer abgeschlossenen Teilmenge eines Banachschen Raumes bezüglich der Differentialgleichung $u' = f(t,u)$, *J. für reine u. angew. Math. 282*(1976), pp. 59-65.

9. Redheffer, R. The theorems of Bony and Brezis on flow-invariant sets, *Amer. Math. Monthly 79(7)*(1972), pp. 740-747.

10. Seifert, G. Positively invariant closed sets for systems of delay differential equations, *J. Diff. Eqs. 22*(1976), pp. 292-304.